Translated and Published for the Agricultural Research Service, United States Department of Agriculture, pursuant to an agreement with the National Science Foundation, Washington, D.C. by Amerind Publishing Co. Pvt. Ltd., 66 Janpath, New Delhi 110001

Translator: R.S. Chakravarthy
General Editor: Dr. V.S. Kothekar

Available from the U.S. Department of Commerce, National Technical Information Service Springfield, Virginia 22161

Printed at Raj Bandhu Industrial Co., New Delhi, and typeset at Oxonian Press Pvt. Ltd., Faridabad
Production: M.L. Gidwani

This monograph presents experimental findings by the author as well as the results of other research here and abroad on the physiology of machine milking of cows. The problems of the physiology of milk letdown in cows with the use of milking machines of different types have been studied with due allowance for the peculiarities of the functioning of the milk glands. Specifically, method is proposed for the physiological evaluation of the effects of using a milking machine. Some tips are offered, such as how to organize machine milking in such a way that the reflex milk letdown is not disturbed.

The book discusses the theoretical and practical aspects of the physiology of milking and increasing milk yield with milking machines.

This monograph is intended for livestock experts, animal breeders, veterinarians, physiologists, engineers and research workers engaged in the study of the problems of machine milking of cows.

Critical notices and requests for a copy of the book may be sent to Urozhai, No. 10, Bolshaya Podvalnaya Street, Kiev 34.

Contents

Introduction

The resolutions of the March 1965 Plenary Session of the Central Committee of the Communist Party of the Soviet Union and the Congress of the Communist Party of the Soviet Union have provided for a number of economic measures directed toward the raising of agricultural production in the country. There remain particularly important problems to be overcome in dairy farming which is still lagging behind. Concurrent with the organization of the fodder base, the complete mechanization of the livestock farm is expected to play an important role in the development of cattle rearing. Machine milking of cows holds a central place in the scheme of rational mechanization of labor consuming processes in dairy farming. Hence the drafting of biologically sound procedures for the introduction of machine milking of cows on the cattle farms of kolkhozes and sovkhozes is of great national importance.

The introduction of milking machines into dairy farming practice requires the reorganization of all technological processes and the standardization of methods of servicing, the maintenance regimen of cows and mechanical facilities on the dairy farm. The mechanical devices used should not only not harm the biological function of the organism but should also stimulate the lactation of the cow since it is on this that the economic result of dairy cattle farming depends. Hence, all technological schemes in dairy farming should have a biological basis.

The applied physiology of machine milking of cows should have a special role to play in the solution of these problems. Experimental analysis of milk letdown showed that the principal reason for unsuccessful trials was incomplete stripping of milk from the udder: after every milking, 2–3 liters of residual milk was held back. At this level of milking within a period of 3–4 weeks a section of the herd being milked with milking machines goes dry, another section is stricken with mastitis, and another suffers a sharp decline in milk production. As a result of this the utilization of milking machines becomes economically unremunerative. Hence the main problem in machine milking lies in the fact that in a limited time the maximum possible quantity of milk should be drawn out of the udder at each milking without harming the health of the animals. For this purpose it is necessary to determine the conditions necessary for the stimulation of reflex milk letdown while using the milking machine. It is, therefore, necessary to estimate the completeness of milk letdown in machine milking by using precise but not unduly cumbersome methods.

The engineering and design component in new milking machines and the development of new technological and organizational schemes of machine milking depend on developments in the field of applied physiology of machine milking. Hence the development and manufacture of milking machines and the technology of machine milking should be subordinated to physiological, zootechnological and hygienic requirements. But the problems of applied physiology of milking of cows by milking machines have, unfortunately, been tackled rather inadequately in our country as in other countries.

Only the most important references in the literature on the subject are cited in this work. They are designed to assist the reader in further study of the problem.

In the review of material for this book valuable criticism and comments were made by Doctor of Biological Sciences Prof. A.D. Sineshchekov, Doctor of Agricultural Sciences Prof. E.A. Arzumanyan and Kandidat of Agricultural Sciences L.V. Kulikov.

Substantial assistance in the field of statistical evaluation of data was rendered by V.A. Rad'ko. In the running of physiological tests with milking machines great help was rendered by Z.F. Zhurba, A.A. Pelekhov, A.G. Eremin and A.A. Korsakov. The author expresses his heartfelt thanks to all his colleagues who took part in discussions on topics treated here or otherwise rendered help in carrying out the experimental work.

Chapter I

Physiology of Milk Letdown

Structure of the udder

The udder of the cow consists of four quarters. The internal cavity of the teat and the large lacteal canals form the cistern of the quarter. At the tip of the teat there is an opening which is closed by contraction of the circular muscles. Thanks to this the milk does not pour out of the udder in the time intervals between milkings. In the walls of the teats and the quarters of the udder is embedded smooth musculature whose special feature is that it may be found either in a state of contraction or relaxation. This special characteristic of smooth musculature has great importance in retaining varying volumes of milk in the udder and in bringing about milk letdown.

Some 20 large lacteal ducts empty into the cistern of each quarter from the alveolar portion of the udder. These ducts branch out upward. Their diameter decreases as their number increases. Each of the smallest lacteal ducts terminates in a vesicle or lacteal alveolus. The diameter of the latter varies from 0.1 to 0.8 mm. It is in the alveoli that the secretion of milk takes place. The inside of the alveolus is lined with secretory cells in which the milk is formed. Secretory cells secrete milk into the cavity of the alveolus, from where the milk flows out through the lacteal ducts.

The alveolus is surrounded by a thick network of minute blood vessels or capillaries, nerve fibers and their terminations or receptors. Through these receptors signals travel from the alveoli to the brain and back. Each alveolus is wrapped in a network of special myoepithelial (stellate) cells which cover the alveolus like a basket. Richardson [149]* was able to obtain very clear photographs of these myoepithelial cells by using thick sections of the alveolar tissue of the udder. When the stellate cells contract the alveolus is compressed, the pressure inside it is raised and as a result its milk content is squeezed out into the duct and milk letdown follows.

*The figure in brackets indicates the serial number of the source of literature cited in the bibliography given at the end of the book.

The arteries which permeate each quarter of the udder supply blood to the alveoli containing nutritive substances out of which the secretory cells manufacture milk. The blood flows out of the udder through the veins. The udder of the cow is connected through a network of blood vessels and nerves to the brain, heart, lungs, stomach, intestines, liver, endocrine glands and other organs [30]. The entire organism of the cow thus takes part in the production of milk.

Secretion of milk

Blood coming from the liver and digestive organs through the capillaries to the alveoli carries nutritive substances, part of which joins the tissue fluid or lymph. The secretory cells of the alveoli are bathed in this fluid. Each cell produces all the constituent substances of milk, i.e. protein (casein, albumen and globulin), fat, milk sugar (lactose) and other substances. From the tissue fluid the secretory cells of the alveoli absorb only the substances required for the formation of milk.

The secretory cell consists of a minute drop of liquid of a very complex structure called protoplasm. The protoplasm of each of these cells is covered by a special enveloping membrane. Milk is formed in the protoplasm and then discharged into the cavity of the alveolus, from where it passes into the lacteal vessel and then to the cistern of the quarters of the udder.

The composition of milk is very complex. The molecules of proteins, milk sugar and mineral substances, all of which are osmotically active, are uniformly dispersed in the water of the milk. Milk fat, which has no osmotic activity, exists in the protoplasm of the secretory cells in the form of groups of very minute droplets or fat globules which enter the alveolar cavity and float about in the water of the milk. As King [91] showed, the fat globules have a very complex structure. The globule consists of droplets of fat covered on the outside by a few layers of proteins, fatty substances and vitamins. All the vitamin A that the milk contains is concentrated in the fat globules. A special feature of the fat globules is the fact that in the small vessels they collect in tiny clusters [197] which cannot proceed through the small vessels. At the same time the water and its contents of smaller dimensions, such as the molecules of proteins and lactose and mineral substances, freely move along the vessels and gradually fill the cisterns of the quarters and teats of the udder. In addition, some of the fat globules are retained in the protoplasm of the secretory cells. They enter the cavity of the alveoli and then the lacteal vessels only at the time of milk letdown [203]. Hence the first batch of milk milked out of the udder contains very little fat, about 1–2%, while the milk derived from the alveoli contains 6–10 and up to 24% fat. A more thorough milking out of the milk from the alveoli hastens the formation of the subsequent lot of milk. Continued disruption of the process of milk letdown may cause a reduction in secretion and lead to the cows going dry.

Unconditioned reflex of milk letdown

The receptors of the teats and skin are stimulated by washing the udder and slipping the milking cups onto the teats and by the repeated pressure on the walls of the teat of the rubber band on the milking cups. These stimuli pass on along the nerves to the spinal column from where reciprocal signals proceed to the smooth muscles of the sphincters of the teats, cisterns and larger lacteal vessels. The smooth muscle fibers relax, as a result of which milk enters the teat cisterns from where it is sucked in by the milking apparatus.

The first phase of reflex milk letdown, which is actuated by the nerves, begins about 5–10 seconds after the start of the stimulation of the teats. The alveolar milk will not have left the alveoli by this time but the milking apparatus draws out milk from the teats. This is due to the fact that by now the degree of contraction of the muscles of the teat sphincters will have come down to one-half or one-third and as a result the sphincters open out much quicker because of the vacuum.

The second phase of reflex milk letdown is connected with the liberation of oxytocin, the milk letdown hormone. Signals from the receptors pass along the afferent nerves (carrying impulses from the receptors to the brain) and reach the posterior lobe of the hypophysis, which then liberates oxytocin into the bloodstream. Oxytocin reaches the udder by way of the blood and brings about the contraction of the myoepithelial cells. By compressing the alveoli these cells cause increased pressure in the alveoli and as a result milk flows into the lacteal vessels and cisterns, from where it is drawn off by the milking apparatus. This second phase of milk letdown begins some 30–60 seconds after the start of stimulation of the teat receptors and lasts 2–6 minutes. Impulses originating in the brain and conveyed along the afferent nerves (carrying impulses from brain to receptors) may also act upon the neuromuscular elements of the udder. A special feature of the milk letdown reflex is that it does not last long and its termination is not dependent on whether the cow is milked or not. These facts must be kept in view in machine milking and the cow should be completely milked out once milking is begun.

The conditioned reflex of milk letdown

I.P. Pavlov [140] laid the foundation of the study of conditioned reflexes. He established that if the giving of food to dogs was associated with sound or a flash of light, then after many repetitions only one ring of the bell or flash of light would bring about salivation in the dog. Something like this takes place in the cow. When the cow gets accustomed to a particular routine even the approach of milking time, the milkmaid's voice, the clatter of the milking equipment and so on may cause milk letdown. The habitual setting in which milking takes place, which the cow gets used to, stimulates the nerve endings (receptors) of eyes and ears. The receptors send signals to the brain cortex from where signals are conveyed to the udder. The smooth muscle fibers of the udder and the teat sphincters relax before milking and this makes for better milk letdown. The milking apparatus draws the milk out of the udder more completely under these conditions.

The study of the possibility of bringing about a conditioned reflex for milk letdown, i.e. by contraction of the myoepithelial cells of the alveoli under the influence of oxytocin, is of considerable importance in the practice of machine milking of cows. Specific experiments [34, 69, 96, 97 and 171] showed that with the help of a conditioned reflex it is possible to get 71.5–95% of the milk from the alveolar part of the udder. These experiments were conducted on the following lines: Before milking the cow a catheter was inserted into the teat of one of the quarters and the cisternal milk was drawn out.* Then the rest of the quarters of the udder were milked to obtain the alveolar milk. After this the conditioned reflex was activated. The proof of the participation of the conditioned reflex for milk letdown was the increase in the volume of the cistern milk at the expense of the alveolar. At present, however, this proof is not sufficient to establish the conditioned reflex letdown of milk in cows. The fact is that after extraction of alveolar milk from the quarters of the udder some residual milk is retained, the residual milk ranging from 50 to 7800 ml [186]. These variations may not affect the cisternal and the alveolar part of the milk, any increase or decrease in which takes place at the expense of the residual milk. Therefore the completeness of evacuation of the udder both in conditioned and unconditioned reflex letdown should be estimated on the basis of a study of the residual milk.

The recovery of the residual milk is associated with the injection of oxytocin, which depresses the impulse converging activity of the cerebral cortex. All this testifies to the fact that for determination of the complex of external conditions activating the conditioned reflex of letdown in cows still more precise methods have to be devised. In order to confirm that under the conditions of an experiment the conditioned reflex letdown of milk takes place, it is necessary to ascertain the oxytocin content in the blood under conditioned as well as unconditioned reflex letdown. Hawker and Roberts [78] reported, for example, that the discharge of oxytocin from

*Terminology adopted for the Symposium on Physiology and Biochemistry of Lactation in 1961 [18] has been used in this book.

neurohypophysis may occur long before milking and the oxytocin content of the blood was high even after milking if the cows remained near the milking parlor.

Then how do we explain the removal of milk from the alveolar part of the udder? We may assume that in response to a conditioned stimulus the removal of milk is the consequence of relaxation of the muscles of the cistern and ducts of all calibers, of distinct fullness of udder at the start of the experiment and of other physiological stimuli. The conditioned reflex of milk letdown must be achieved through the same afferent paths as unconditioned reflexes. That means that impulses starting from the cortex of the brain under the influence of the conditioned stimuli would be directed to two destinations—through the spinal column to the smooth muscles of the tissue of the udder inducing relaxation, and secondly, through the stalk of the hypophysis to the hypophysis, bringing about release of oxytocin which, reaching the myoepithelial cells by way of the bloodstream, brings about contraction. In order to prove that when a conditioned reflex is activated everything in fact proceeds in a set pattern the evidence would have to be more convincing. This does not imply that a conditioned reflex milk letdown should not be developed. In the natural conditions of the cattle yard a conditioned reflex of milk letdown from the alveoli is probably activated. Our task now is to assess this reflex and to reproduce it under experimental conditions, having determined the combination of external conditions required for this.

Inhibition of milk letdown

The milk letdown reflex of cows is very easily retarded by external stimuli arising from disruption of the usual arrangements for milking. Hence the problem of inhibition of milk letdown assumes great practical importance. It is supposed that inhibition of milk letdown may be related to the release of adrenalin from the cortex of the adrenal glands [29, 35 and 134], which brings about the contraction of the smaller vessels of the udder through the action of the afferent nerves by delaying the release of oxytocin [147] and interfering with the conditioned reflexes of milk letdown [20]. The problem of countering inhibition of the milk letdown of cows is far from a practical solution. So far it is not known under what conditions a particular kind of inhibition appears, such as those related to adrenalin, afferent nerves, withholding of oxytocin and interruption of conditioned reflexes on milk letdown. The correlation of these factors in the overall phenomenon of inhibition of milk letdown has not been studied and the quantitative characteristics of letdown are not known. Ducembin [56], for example, determined milk letdown from the variations in the reflex milk letdown (in response to milking for 30 seconds) and the volume of alveolar milk stripped. But this author did not, unfortunately, study the residual milk.

Oxytocin, the milk letdown hormone

The neuroendocrinal theory of lactation has now been fairly well established. According to this theory the nervous system and hormone oxytocin or a complex of hormones participate in reflex milk letdown. According to modern classification [37] oxytocin belongs to the group of kinetic hormones that bring about rapid and repeated action of effector organs. It is supposed [79] that oxytocin is formed in the paraventricular and partly in the supraoptical core of the hypothalamus from where the neurosecretion is conveyed along the fibers of the hypothalamo-hypophysial bundle to the neurohypophysis (posterior pituitary) where it is partly stored and partly carried away by the bloodstream. The neurohypophysis of cattle contains some 198–324 international units (IU) of oxytocin [106]. The structure of oxytocin has been studied thoroughly [28]. It is a cyclic octapeptide with a molecular weight of 1007 and comprises eight amino acids. Oxytocin has been synthesized in the laboratory.

It is well known that the milk letdown reflex is of very short duration. The study of the duration of the milk letdown reflex involves the estimation of the oxytocin content in the blood, lymph and plasma of various organs. Before milking and at milking time cows have shown 380 milliunits/ml oxytocin content in the jugular blood and after milking 80 milliunits/ml [78]. (One international unit contains 100 milliunits of oxytocin.) In similar experiments [63] with nine sheep, before milk letdown oxytocin content of the blood was 2.6–7.8, at the time of letdown 21.0–114.0, and two minutes after letdown 2.5–7.7 milliunits/ml. In one ewe the activity of the blood was determined every 10 seconds at the time of suckling the lamb. Before suckling the oxytocin content of the blood was 5, in the first 10 seconds of suckling 39, then 28, 15 and back to 5 milliunits/ml. A high concentration of oxytocin in the blood of the ewe at the time of suckling was maintained through 30 seconds. This experiment explains the brief duration of the milk letdown reflex in domestic animals. Rapid withdrawal of oxytocin from the blood takes place independently of whether the animal has been milked or not. This fact is of importance for the organization of a regular system of machine milking of cows.

What are the factors that regulate the rapid disappearance of oxytocin from the blood and the internal medium of the effector organs on which it acts? Formerly it was thought that in farm animals oxytocin breaks down in the blood. But this was not corroborated. It was observed that the udder, liver, spleen and other organs rapidly deactivate oxytocin. One gram of the homogenate of the liver of a cow, for example, deactivates 2 IU of oxytocin in some 15 minutes, while the blood plasma of cattle and other animals did not break down oxytocin. This shows that the blood of farm animals serves only to transport oxytocin to effector organs. Oxytocin is deactivated by enzymes: in the kidney, for example, deactivation takes place by enzymes such as glutaminase and asperginase. In the udder, oxytocin brings about contraction of the myoepithelial cells that compress the alveoli and cause them to empty. The molecule of oxytocin is broken down by enzymes of the myoepithelial cells during this process.

Thus the tissue fluid of organs rich in enzymes is the place where the oxytocin is deactivated; the liver, kidneys, spleen, lacteal glands, skeletal and smooth muscles, etc. The blood plasma of the animals does not deactivate the oxytocin. But since the blood circulates in the body quite fast (blood circulates through the entire body of the cow in 20–30 seconds) the oxytocin that gets into the tissue fluid of many organs is rapidly disintegrated by enzymes and the concentration of this hormone in the blood falls rapidly. In experiments with sheep [116] an injection of 0.25 IU of oxytocin into the jugular vein raised the blood pressure by 8 cm of water column while an injection into the mesenteric vein did not change the pressure. In the latter case the oxytocin, which had to pass through the liver, was deactivated.

In experiments on laboratory animals [172] when oxytocin was injected into the jugular vein at doses of 0.3–1.2 milliunits/ml milk letdown set in, and (this happened) after injecting doses 50 times larger into the saphenous artery (16–20 milliunits/ml). This showed that oxytocin introduced into the artery did not enter the milk gland but into the tissue fluid of other organs where it was deactivated by the enzymes. These data show that the method of injection of oxytocin into the tissue fluid of organs other than milk glands is unreliable and, therefore, of no use in the study of lactation of the cow. It is unnecessary to introduce oxytocin into the jugular vein for the study of lactation under normal conditions. But unfortunately the latest studies on oxytocin have not been taken into account in Tverskoi's extensive research [177, 178, 179 and 180] on the role of the nervous system in the regulation of milk secretion and the milk letdown reflex. He characterizes the physiological processes according to the quantity of milk obtained from the udder of goats after intramuscular injections of 3 IU of oxytocin. But this dosage of hormone is insufficient for extraction of all the milk from the udder of a goat. For example, Semjan [156] showed that even in response to 6–10 IU of oxytocin in subcutaneous injection in sheep only 73% of the residual milk could be drawn out. The rest was obtained with the help of a second injection. It was not for nothing that Cowie and Tindall [42] had to give an intravenous injection of oxytocin in doses of 10–30 IU to induce milk letdown after the removal of the pituitary body. Therefore, the conclusion arrived at by Tverskoi and other workers, who employed subcutaneous or intramuscular injections of oxytocin, requires more accurate experimental proof. The role of oxytocin is not limited to its participation in milk letdown. It also has a stimulating effect on milk secretion [190].

The mechanism of milk letdown

A special characteristic of the milk letdown reflex in cows is its short duration. It is known that of the total quantity of milk accumulating in the udder before milking only one-half is obtained in milk letdown, i.e. conditioned response of the cow. If before milking a catheter is introduced into the cow's teat approximately half of the milk flows out of the quarter of the udder without milk letdown. This is the cisternal milk which is distributed in the teat cisterns and the larger lacteal vessels. The milk from the alveoli may be obtained by milk letdown, which sets in 30–60 seconds after the beginning of the stimulation of the teat receptors. This is the latent period of the milk letdown reflex. Milk which is drawn out of the udder during milk letdown is known as alveolar milk. This terminology was first proposed by Zaks [204]. Milk letdown in the cow starts practically simultaneously in all four quarters of the udder and the duration of the alveolar milk flow is about 3 minutes on an average [182]. It was also found [183] that when each of 12 cows were milked by two milkmaids the duration of milking was on an average 2 min 57 sec. One milkmaid milked out each cow in 6 minutes on an average and when the completeness of milking was estimated it turned out that 6.6 liters of milk was obtained in this milking while 2.0 liters or 23.2% of total milk was retained in the udder. But if the same cows were milked out in 2 min 57 sec 7.1 liters was obtained and 1.1 liters or 13.4% of the total remained in the udder. Thus rapid milking not only results in saving milking in time but also brings about a fuller milking out of the udder. This finding is of great importance for the practice of machine milking of cows.

Why is there a fuller evacuation of milk in rapid milking? What is the reason for this phenomenon? We may suppose that, with the available methods of hand and machine milking, during the latent period of reflex milk letdown (30–60 sec) the cisternal milk cannot be milked out and the reflex begins with a full cistern. In spite of the fact that the smooth muscles of the cistern relax the latter cannot hold all the milk coming from the alveoli. This may hinder the milk letdown and lead to an increase in the residual milk. To prove this hypothesis some special experiments were conducted with five cows. In all 95 trials were run in four series. The experiment was carried out as follows: Before milking, the cistern milk was drawn out of one or more quarters of the udder by means of a metallic catheter. After this the milkmaid milked out alternately the rear quarters of the udder for 3–4 minutes and then the front quarters for 2–3 minutes into a milk pail with four partitions. Eighteen IU of oxytocin* was then injected into the cow's jugular vein and the residual milk milked out from each of the quarters. The interval allowed between the successive experimental milkings was about 10–12 hrs. The completeness of milk letdown was estimated like this: The total quantity of milk and fat obtained from the quarters was assumed to be 100, and the cisternal, alveolar and residual milk were expressed as percentages of total milk and fat obtained from the quarters.

In the first series of experiments, the cisternal milk was removed before milking from the right rear quarter and the rear quarters were then milked out to begin with. Results of one of the experiments on the cow Angar are given in Table 1. The average of 35 trials in this series showed that in

TABLE 1. EFFECT OF REMOVAL OF MILK FROM THE CISTERN CAVITY BEFORE MILKING ON THE COMPLETENESS OF EVACUATION OF MILK FROM THE FOUR QUARTERS

Characteristics	The four quarters of the udder			
	Right front	Left front	Right rear	Left rear
Milk yield (ml)	930	850	670*/1,560	2,270
As % of milk of quarters	71.5	72.0	18.1/75.6	87.3
% of fat	2.1	2.1	0.4/4.5	2.4
Weight of fat (g)	19.5	17.6	2.7/70.2	54.5
As % of fat of quarters	44.5	46.0	3.1/80.2	66.3
Residual milk (ml)	370	330	150	330
As % of milk of quarters	28.5	28.0	6.3	12.7
% of fat	6.6	6.8	9.6	8.4
Weight of fat (g)	24.4	22.4	14.4	27.7
As % of fat of quarters	55.5	56.0	16.7	33.7

*In the case of the right rear quarter the numerator indicates the cisternal milk and the denominator the alveolar milk.

*For this purpose it is better to use synthetic oxytocin which is supplied from Hungary on order to "Soyuzzoovetsnaba." In the absence of oxytocin, pituitrin "M" or pituitrin "R" may be used.

the right rear quarter (figures in percent) 10.6 residual milk and 28.1 fat were retained, while in the left rear quarter the corresponding figures were 17.6 and 41.6 in the right front quarter 25.0 and 55.6 and in the left front quarter 23.7 and 51.9.

In the second series of experiments the cisternal milk was removed before milking from right and left rear quarters of the udder, which were then milked out to begin with; then the front quarters were milked. The results of this experiment showed that completeness of removal of milk from the right and left rear quarters was similar.

TABLE 2. EFFECT OF REMOVAL OF CISTERNAL MILK BEFORE MILKING FROM THREE QUARTERS ON THE COMPLETENESS OF EVACUATION OF THE QUARTERS OF THE UDDER

Characteristics	The four quarters of the udder			
	Right front	Left front	Right rear	Left rear
Milk yield (ml)	640/870	560/970	1240/1000	2300
As % of milk of quarters	38.0/51.4	32.7/56.8	47.3/37.2	80.4
% of fat	0.3/4.4	0.5/4.9	0.5/4.8	1.9
Weight of fat (g)	1.9/38.3	2.8/47.5	6.2/48.0	43.7
As % of fat of quarters	3.3/68.5	4.1/70.7	6.3/49.3	43.6
Residual milk (ml)	180	180	380	560
As % of milk of quarters	10.6	10.5	14.5	19.6
% of fat	9.2	9.4	11.4	10.0
Weight of fat (g)	16.6	16.9	43.3	56.0
As % of fat of quarters	28.2	25.2	44.4	56.4

In the third series of experiments the cisternal milk was removed from the right and left front and right rear quarters of the udder. After this the front quarters were milked out in the first 2–3 minutes, then the rear quarters were milked out. The results of this experiment on the cow Angar are given in Table 2. According to the average of 20 experiments run in this series there was a retention (in percent) of milk 8.7 and fat 20.5, in the right front quarter, milk 9.5 and fat 22.5 in the left front quarter, milk

18.5 and fat 40.3 in the right rear quarter and milk 29.8 and fat 60.1 in the left rear quarter.

The milk letdown indices of the right rear quarter are of particular interest. In spite of the fact that milk letdown in this quarter was to an emptied cistern there were more residual milk and fat in the alveoli than in the front quarters. These data show that milking which begins 2–3 minutes after reflex milk letdown started, even in conditions where cisternal milk had been removed from the quarters before milking, leads to retardation of the flow of milk from the alveoli and to an increase in residual milk and fat. Relaxation of smooth muscles of the cisternal cavity of the udder cannot fully obviate the inhibition of milk letdown in cases where milk flowing out of the alveoli is not removed immediately by milking. It is difficult, therefore, to agree with Zaks [202] that the negative effect of extended 20 min milking is completely averted if the cisterns of the udder are previously emptied. He did not take into consideration the residual milk and did not have the opportunity to specify accurately the effect of extended milking on the completeness of milk letdown of the quarters of the udder from which the cisternal milk has been previously removed.

In the fourth series of experiments the cisternal milk was drained out from all four quarters and then the alveolar and residual milk was milked out. The rear quarters were milked out in the first instance. Of 20 experiments in this series residual milk 9.3% and fat 17.5% was retained on an average in the right quarter; the figures in the left front quarter were 11.4 and 19.1; in the right rear quarter 4.5 and 8.4; and in the left rear quarter 5.0 and 9.5%. As we can see, milk letdown from alveoli to the evacuated cisterns is more effective when the milking out of alveolar milk entering into the cisterns is done at once.

Results of the above experiments show that the peak of contraction of the myoepithelial cells of the alveoli lasts in all for some 1–2 min and this is followed by relaxation. Cisternal milk, when it is not removed by milking before commencement of milk letdown, inhibits the passage of milk out of the alveoli. It is quite probable that slow removal of cisternal milk during milking is one of the causes for the holding back of residual milk in the udder of the cow.

Repeated milk letdown in the cow

The yield of milk obtained in milking is used as an index in the study of lactation. But since residual milk, which is subject to variations, is held back in the udder after milking, it is necessary to take into account not only the milk yield but also the residual milk for any precise characterization of the process in question. To draw out the residual milk from the udder after milking oxytocin is injected, or, in other words, a repeated or second letdown of milk is brought about. In the study of the second letdown of milk in a cow the dose of oxytocin used is very important. In various experiments the dose of oxytocin used to draw out residual milk has varied between 3 and 60 IU. This has led to the accumulation in the literature of incredibly variegated and mutually contradictory data concerning the secretory processes of milk letdown and other problems of lactation. Moreover, Hammond [76] formulated the basic procedural conditions for evaluation of the different doses of extract of the posterior lobe of the pituitary (the individual hormones were not yet known at the time), which were then used in all subsequent studies to ascertain the doses of oxytocin necessary to induce a repeated letdown of milk in cows.

But this method does not take into account the kinetic nature of oxytocin. It is known that oxytocin is able to induce repeated contractions of the myoepithelial cells of the alveoli. Therefore the completeness of evacuation of the udder after an injection of the dose of oxytocin may be controlled with the aid of repeated injection of an effective dose of this hormone. If milk is present in the alveoli it will be squeezed out into the cavities of the lacteal vessels and cisterns, from where it may be removed by massage of the udder and hand milking. The correlation of residual milk obtained with the first (trial) dose of oxytocin with the volume of residual milk obtained with the second (control) dose of oxytocin will give a correct indication of the effect of doses of oxytocin in repeated letdowns of milk in the cow. It makes possible exclusion of the effect of the variability factor of residual milk on the result of the evaluation of the oxytocin doses. These premises were made the basis of our studies to determine the doses of oxytocin necessary to induce a repetition of the letdown and complete withdrawal of residual milk from the udder after the usual hand or machine milking of the cow.

The experiments were conducted on five cows. The experiments were run as follows: After hand milking of the cow the test dose of oxytocin was injected into the jugular vein and immediately the residual milk was milked out. After the termination of the letdown response to the milking stimulus induced through the jugular vein, the control dose of the oxytocin (24 IU) was injected and the residual milk was again milked out. The period during which the cow was exposed to the experiment did not exceed 25–30 minutes, which was enough to exclude the effect on the result of the experiment of any milk being secreted then. This factor was not, for example, considered by Johansson [83] in his experiments and this led him to a faulty conclusion.

Research was carried out in the first half of lactation in 135 experiments conducted in 10 series over 108 days. Injections of oxytocin were given once in four days in the morning milking with an interval of 10 hrs between milkings. Doses of 3–24 IU of oxytocin were used. The series of

TABLE 3. EFFECT OF OXYTOCIN DOSES ON REPEATED LETDOWN OF MILK IN THE COW

Oxytocin doses (I.U.)		Single milking (ml)	Residual milk			Residual milk as % of total volume of residual milk in udder		
			After 1st injection of test dose		After 2nd injection of test dose (ml)	By 1st injection		By 2nd injection
Test	Control		ml	As % of milk yield+residual milk after 1st injection		Average	Fluctuation	
18	24	8,240	1,380	14.3	13	99.1	96.3–100	0.9
3	24	8,500	209	2.3	1,246	14.3	0–100	85.7
66	24	7,840	740	8.6	800	48.1	8.8–100	51.9
9	24	7,240	1,370	16.0	513	72.8	50.0–100	27.2
12	24	7,000	1,066	13.0	250	81.3	50.0–100	18.7
15	24	7,000	1,250	15.0	128	90.7	66.7–100	9.3
18	24	6,590	770	10.4	12	98.5	83.4–100	1.5
21	24	6,086	1,063	15.0	20	98.2	82.7–100	1.8
24	24	6,400	1,013	14.0	8	99.2	95.7–100	0.8

experiments where a dose of 18 IU was used was run twice. Results of these experiments are given in Table 3.

From Table 3 we find that it is not possible to ascertain the dose of hormone required for complete withdrawal of residual milk from the udder by looking at the quantity of residual milk obtained in response to the injection of the test doses of oxytocin. According to these indices doses of 9–24 IU of oxytocin have almost the same effect, which we find is in the range of 13–16%. After the test doses still more residual milk can be drawn out with the help of the control dose of hormone and the correlation of the latter quantity of milk with that obtained from the first injection of the test dose of oxytocin affords a proper indication on the effect of the doses of oxytocin on complete evacuation of residual milk from the udder.

It is not possible to draw out all the residual milk from the udder of the cow with 3–15 IU of oxytocin. Increasing the dose to 18 IU makes it possible to drain out of the udder practically all the residual milk in the next thorough milking. A control injection of 24 IU after 18 IU does not increase residual milk. Doses of 21 and 24 IU as compared to 18 IU have approximately the same effect. These data show that with the help of injections of 18 IU of oxytocin it is possible to obtain all the residual milk detained in the alveoli of the cow's udder.

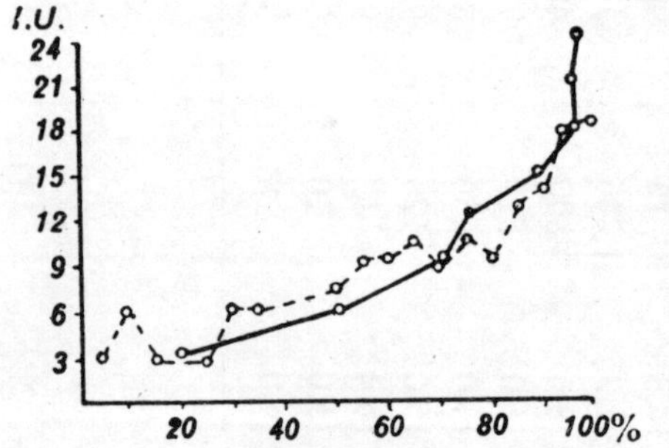

Fig. 1. Regressive relation of completeness of repeated milk letdown of cows to magnitude of oxytocin doses: empirical curves of regression. Oxytocin IU in the ordinate and completeness of milk letdown (%) in the abscissa.

In experiments conducted to determine the optimal doses of oxytocin for repeated milk letdown in cows we used pituitrin "R", containing 3 IU oxytocin per ml. Pituitrin "R" also contains the hormone vasopressin which has some oxytocic activity. Therefore the optimal dose of synthetic oxytocin for drawing out residual milk may possibly be somewhat higher than that of oxytocin in the form of an extract of pituitrin.

The results obtained on the basis of arithmetical calculations confirm statistical analysis. The correlation coefficient between the oxytocin doses and completeness of repeated milk letdown is 0.79. The connection between the two indices is high and reliable. The empirical lines of regression show (Fig. 1) that between the magnitude of doses of oxytocin and the completeness of evaluation of milk in repeated milk letdown there is an intimate relationship—the regression lines almost coincide—but this regressive relationship has a curvilinear character.

Chapter II

Role of Residual Milk in Udder

Residual milk and completeness of letdown

In the practice of machine milking of cows it is essential to know to what point the udder can be emptied of milk during milking. The fact is that increased milk yield at each milking is dependent on complete milking out of the udder. Azimov [12], for example, writing on this problem, said that even after a thorough milking out of the udder there is always some residual milk or strippings left, about a liter or more. But how can this be? Johansson [83], who in 103 trials on cows injected into the lacteal vein 25 IU of oxytocin after machine milking, found that 0.34 kg to 1.64 kg of residual milk remained. In one trial the milking machine obtained a yield of 93.9% of the milk. Koshi and Petersen [101], who conducted 1,327 trials on 62 cows, found variations in residual milk ranging from 44 to 6,024 g. Lauener [111], who ran 40 experiments on 20 cows, obtained some very important results. Residual milk was drawn out by means of intravenous injections of 20 IU of oxytocin. Variations in residual milk ranged from 0 to 7 liters in the experiments. In four experiments there was no milk after machine milking but in seven experiments some 0.1–0.2 liter was recovered. These experiments showed that all the milk could be drawn out of the udder by the usual milking procedure. It is natural that for a correct understanding of the milk letdown phenomenon and the evaluation of practical approaches to the study of factors stimulating this phenomenon it is necessary to know more about the mechanism underlying the variation in residual milk in the udder of the cow.

We studied milk letdown in 189 cows [186] and ran in all 990 experiments along these lines: After hand or machine milking we injected 18 IU of oxytocin by way of the jugular vein and then residual milk was milked out by hand. The second letdown of milk in the cows set in after an average of 17 seconds. The interval varied from 2 to 60 seconds. The character of the distribution of residual milk in relation to the volume and number of experiments is shown in Fig. 2. From the graph it may be seen that the residual milk is distributed in narrow limits in which half of the total experiments have up to one liter of residual milk. According to the data of the 990 trials the milking varied from 0.1 to 11.7 liters and the residual milk from 50 to 7,800 ml. To what extent were these variations interrelated? Since the investigation of this problem is of significance for a correct understanding of the milk letdown phenomenon we have grouped the trials into 11 classes in relation to total volume of milk in the udder and calculated within the limits of each class the correlation coefficient between the milk yield and residual milk. The results of this study are presented in Table 4.

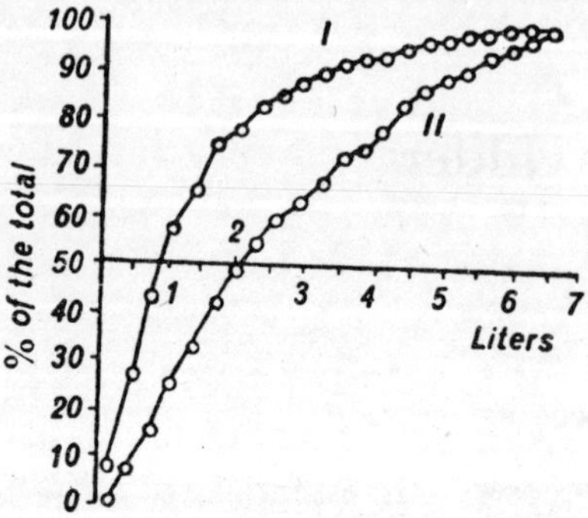

Fig. 2. Probability distribution of residual milk.

I—Calculated distribution; II—Volumetric distribution.

It can be seen that the correlation coefficient between foremilk and residual milk shows an inverse relation with a greater degree of reliability. Therefore the yield increase following complete milking out is always accompanied by a decrease in residual milk. On the other hand, increase in residual milk always entails a diminution in milk yield. It is necessary to emphasize that in cows there is no established level of

TABLE 4. CORRELATION COEFFICIENT BETWEEN FOREMILK AND RESIDUAL MILK IN COWS

Total volume of milk in udder (liters)	No. of trials	Correlation coefficient	Error of correlation coefficient	Standard deviation
3	35	−0.910	0.072	12.5
4	67	−0.912	0.051	17.9
5	136	−0.970	0.005	191.8
6	215	−0.999	0	—
7	176	−0.905	0.013	66.0
8	131	−0.925	0.012	73.5
9	98	−0.975	0.002	42.7
10	62	−0.970	0.031	30.6
11	35	−0.698	0.124	5.62
12	25	−0.717	0.145	4.94
13	8	−0.988	0.060	16.3

emptying the udder, no indications or sign that demarcate foremilk and residual milk. The results obtained show that residual milk in the udder may not be constant from the beginning.

Out of 990 trials in 35 or 3.5% the udder was almost completely emptied by the usual procedure. In these trials, after hand or machine milking the oxytocin injection extracted in all some 50–200 ml of residual milk. In approximately 50% of the 35 trials the entire quantity of residual milk has been retained in one or two quarters: the other quarters were empty. From these quarters even after a vigorous massage and injection of oxytocin not 10 ml of milk could be drawn. We may ask whether there was actually no residual milk at all in the udder in these cases. The fact is that intravenous injection of oxytocin involves the excitation of the cow and painful reactions which, as is known, activate the sympathetico-adrenal system of the reflex letdown of milk. In regard to this we may assume that in those experiments where no residual milk was obtained or where its volume was insignificant it was not a case of no residual milk in the udder but the blocking of the physiological activity of the exogenous oxytocin by adrenalin as a result of which the movement of the residual milk out of the alveoli was impossible. To verify this assumption cows were given repeated injections of oxytocin. After milking out the residual milk from the first injection of the cow 18 IU of oxytocin was injected intravenously and milking was begun immediately. Results of these experiments showed that all the milk from the cow's udder had been milked out in the usual milking and a two-stage injection of oxytocin did not produce any changes in this state of affairs. Evidently the reflex letdown in conjunction with the stimulus of rapid milking provides conditions under which the udder can be completely emptied of milk. Results obtained allow us to presume that the normal reaction of the cow to being milked promotes the release of oxytocin into the blood from the neurohypophysis in quantities sufficient for active milk letdown and removal of all the milk from the alveoli of the udder. The data from these trials show that complete emptying of milk from the udder at milking is a normal physiological phenomenon and the presence of residual milk in the udder may be considered to be due to inhibition of the cow's milk letdown reflex. Correlation of foremilk and residual milk indicates the degree of inhibition of letdown in the animal. Residual milk in the cow's udder is not a constant phenomenon, its magnitude varies, its volume is subject to sharp changes and it is amenable to recovery by the usual method of milking.

By completeness of milk letdown we understand the extent of draining (of milk) of the udder during milking by conventional methods of milking. The completeness of milk letdown can be expressed as a percentage of the entire milk produced in the udder, including even the residual milk. The milk yield, foremilk and residual milk is expressed as a percentage of the total volume of milk accumulated in the udder by the time of milking.

The quantity of milk recovered from the cow's udder almost fully by the usual method of milking out is very little, but the data on this have great practical significance in machine milking since they indicate that with the present methods of milking conditions exist under which it may be possible to recover all the milk from the cow's udder. According to our findings the following factors play a part in bringing about prompt, complete evacuation of milk from the alveoli under conditions of natural milk letdown: the amount of oxytocin released into the blood from the neurohypophysis in response to stimulation of the teats, the activity and duration of the contractile reaction of the myoepithelia of the alveoli in response to oxytocin, the magnitude and duration of reflex relaxation of the (lacteal) ducts, cisterns and sphincters of the udder and finally the external setting for milking which generates the conditioned reflex reaction of the cow to being milked and intensifies unconditioned reflex letdown. Hence the improvement of the milking machine, method of milking installation and external conditions, all of which promote complete evacuation of milk from the udder at each milking, have real physiological significance.

Cows differ with regard to completeness of milk letdown. Animals which retain a large quantity of residual milk in the udder have a weakly expressed lactation function and dry up rapidly [20, 59]. It is in this context that the study of the impact of the individual peculiarities of cows on their ability to evacuate milk from the udder at milking time assumes great practical significance. An analysis of these data is presented in Table 5. This shows for each cow average data of 18 trials carried out at the same interval and milking. Oxytocin was injected once every three days at one of the milkings over a two-month period. As may be seen from the table, on an average 2.5 times more residual milk was retained in the udder of the

TABLE 5. VARIATIONS IN FOREMILK AND RESIDUAL MILK IN COWS WITH DIFFERING LEVELS OF EVACUATION OF UDDER (MACHINE MILKING)

Factor	Dil'da	Anuska
Foremilk (liters)	6.4	6.3
Fluctuation	3.1–11.4	1.9–10.4
Residual milk (liters)	3.2	1.3
Fluctuation	0.4–6.1	0.2–3.7
Variation coefficient	65.75±10.95	83.30±13.90
Residual milk as % of total milk in udder	32.6	17.1

cow Dil'da than in that of the cow Anuska. Yet this circumstance had no effect on the variation in the residual milk of the animals. In both cows the fairly sharply expressed variations in foremilk yield and residual milk are wide enough to establish that the individual peculiarities of the two animals, whatever their significance, could not be the cause of retention of residual milk in the udder. In Dil'da completeness of evacuation of milk from the udder fluctuated in the range 34.1 to 95.4 percent; in Anuska 37.3 to 96.3 percent. So in cows differences in their ability to evacuate the milk from their udder more or less completely at usual milking time cannot serve as the starting point to prove that individual peculiarities of cows govern the retention of milk in the cow's udder. Apparently for all cows there exist certain general conditions under which all or almost all the milk can be milked out of the udder.

Effect of age of cow on completeness of letdown

Among the factors that may contribute to occasional holding back of milk (residual) in the cow's udder, the age of the animal may be considered as one [11]. However, Johansson [83] did not notice any influence of age on the volume of residual milk in cows. He observed only that there might be a tendency for age to play a part. Koshi and Petersen [101] found the correlation coefficient between the age of the cow and percentage values of residual milk equal to 0.18, which points to the absence of any relation between these two factors. As the determination of the role of the cows age among the complex of other factors governing the retention of residual milk in the udder is of great importance for standardizing the technology of machine milking of cows, we processed and distributed the data of 990 trails on 189 cows in 12 lactations. The cows whose data were analyzed ranged from 2½ to 16 years of age. The results of this investigation are presented in Table 6.

As may be seen from this table, wide variations in residual milk are observed in cows of all ages and the degree of evacuation of milk from the udder is roughly the same in young and old cows.

The correlation coefficient between age and volume of residual milk was equal to $+0.239$. If we assume that the relation between the factors may be expressed as the square of the correlation coefficient it appears that a relation between age and volume of residual milk can be established in only 5.7% of the experiments. This implies that the age of the cow has no effect on the volume of the residual milk. On the other hand, the magnitude of the total volume of milk in the udder or the level of productivity of the cow may have an effect on the relation between the age of the cow and the volume of residual milk.

To establish the degree of influence of this factor we calculated the

TABLE 6. CHANGES IN VOLUME OF RESIDUAL MILK IN COWS OF DIFFERENT AGES

Lactation of cow	Number of cows	Number of trials	Fore-milk (liters)	Residual milk (liters)		
				Average	Fluctuation	As % of total milk in udder
1	20	150	3.12	1.17	0.1–5.1	27.2
2	38	217	5.46	1.23	0.2–6.0	18.2
3	21	164	5.55	1.10	0.2–5.6	16.5
4	37	93	5.58	1.41	0.2–6.1	21.4
5	8	61	6.00	1.90	0.1–6.2	24.0
6	8	38	4.39	2.61	0.3–6.5	37.9
7	12	60	4.76	2.49	0.3–5.1	34.3
8	9	45	4.75	2.67	0.5–6.3	36.1
9	13	43	5.70	1.40	0.3–3.0	20.0
10	12	39	5.48	1.67	0.2–4.5	23.3
11	6	42	5.40	1.76	0.3–6.0	24.5
12	5	38	5.34	1.89	0.3–7.8	26.0

partial correlation coefficient between the total volume of milk in the udder, the age of the cow and the residual milk. This turned out to be +0.147. The results of this analysis show that in each age group of cows there is no biologically inherent faculty to hold back in the udder a regular large or small quantity of residual milk. Old cows may be as completely milked out by a milking machine as young cows.

Effect of level of productivity on completeness of milk letdown

It is said [118, 175] that with an increase in the milk yield of cows the quantity of residual milk in the udder also increases. According to the data of Koshi and Petersen [101] it depends on the level of productivity but completeness of evacuation of the udder or the percentage of residual milk does not depend on the level of productivity of the cows. The correlation coefficient between these factors is, according to their data, 0.02. Perhaps the conclusion arrived at by these authors was to a considerable degree the result of incomplete repeated milk letdown, seeing that they employed only 10 IU of oxytocin to draw out the residual milk.

The findings from our trials have been briefly presented in Table 7 in which 990 experiments have been arranged according to the total volume of milk in the udder of the cows. As can be seen from the table the quantity of residual milk increases with the increase of the total volume of milk in the udder but the completeness of the letdown at all levels of productivity is practically the same. These data indicate that the level of productivity does not exert any effect on the completeness of evacuation of milk from the udder during milking.

TABLE 7. VARIATION IN RESIDUAL MILK IN THE UDDER IN RELATION TO PRODUCTIVITY OF COWS

Total milk in the udder (liters)	No. of trials	Single milking yield (liters)	Residual milk (liters)		
			Mean	Fluctuation	As % of total milk in udder
Up to 3	35	1.94	0.75	0.1–2.3	28.0
4	67	2.77	0.86	0.1–3.1	23.6
5	134	3.76	0.90	0.1–4.5	19.3
6	215	4.35	1.24	0.2–5.3	22.1
7	176	4.90	1.76	0.1–6.0	26.4
8	131	5.95	1.63	0.2–5.6	21.5
9	98	6.53	1.98	0.2–6.3	23.2
10	62	7.19	2.36	0.5–6.0	24.7
11	35	8.00	2.54	0.4–7.8	24.1
12 and above	37	8.90	3.02	1.0–6.5	25.3

The correlation coefficient between total volume of milk in the udder and residual milk is equal to +0.415. In other words, with an increase in total volume of milk in the udder the increase in the quantity of residual milk was observed only in 17.2% of the experiments. Thus the relation between these factors, established on the basis of average arithmetical data, is not confirmed by statistical analysis. The coefficient regression between residual milk and total volume of milk in the udder is 0.248. This indicates that with an increase in total volume of milk by one liter, the quantity of residual milk increases by 248 ml. But this does not agree with the experimental data since the volume of residual milk following an increase in total volume of milk in the udder by one liter fluctuates between +520 and −70 ml. In view of the advantage of the graphic method of regression analysis, empirical lines of regression were constructed and with the help of formulation of levels of regression the theoretical (expected) lines of regression were calculated. The results are presented in Fig. 3. The regression lines on the graph show that the relation between total volume of milk in the udder and residual milk is very distant: the discrepancy between the lines of regression is quite clear. These data show that total volume of milk in the udder or the level of productivity of the cow do not govern the fluctuations in residual milk.

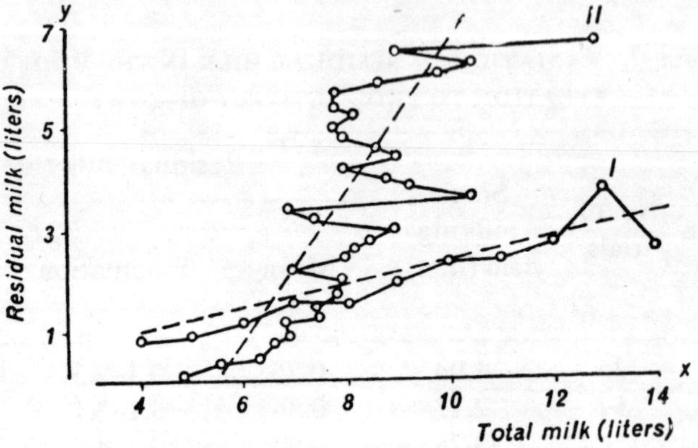

Fig. 3. Regression relation between total volume of milk in the udder (x) *and residual milk* (y).

I—y as a function of x; II—x as a function of y. Empirical lines of regression zigzag, theoretical lines of regression are straight.

Important results were obtained in arriving at the theoretical lines of regression. On choosing the initial value of the total volume of milk for regression levels this figure was found to be 5.5 liters. This is taken from the correlation grid on the following grounds: On analysis of the first two variants of total volume of milk of the correlated grid it appeared that if there was 5.5 liters of milk in the udder at the start of milking it could be almost completely milked out with the existing technique. In this case the residual milk varied from 0 to 0.3 liter. Since milk letdown in the cow usually lasts 4–5 minutes and the rate of milking out by hand machine fluctuates around about one liter a minute it follows that when the volume of milk in the udder is up to 5.5 liters the inhibition of letdown does not set in during the act of milking and the udder is almost completely evacuated. Therefore for complete milking out of the udder of cows of higher productivity it is desirable to raise the speed of machine milking accordingly. This, again, is bound up with improvement of milking machines. So in our experiments we did not find any facts on the basis of which we could suppose that highly productive cows cannot be completely milked out during normal milking. There are conditions in which it becomes possible to evacuate milk fully from the udder of cows of dissimilar productivity. The study of this group of conditions is one of the most important problems of the applied physiology of machine milking of cows.

The influence of residual milk on butterfat yield of cows

In cows that give the same quantity of milk yield in a single or daily milking and are maintained under the same feeding regime and management the fat content in milk may be doubled from 3 to 4 percent to 6 to 9 percent [31, 82, 167]. It is also known that the cow Fascination of Guernsey stock gave 5,946 kg of milk with 10.58% fat content in one of her seven lactations while the average fat content of the milk over her seven lactations was 6.16% [194]. Research scientists have not been able to explain this phenomenon since factors like the influence of the thyroid gland, pituitary, rumen and other organs, specifically effective feeds and feeding regime, exercise and massage of the udder had negligible influence and lay in the range of 0.2–0.8%.

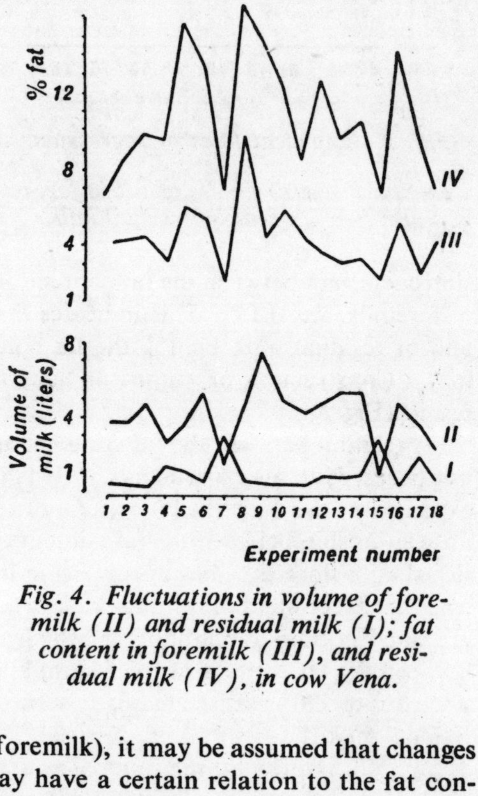

Fig. 4. Fluctuations in volume of foremilk (II) and residual milk (I); fat content in foremilk (III), and residual milk (IV), in cow Vena.

As residual milk is rich in fat compared to foremilk (drawn in the normal milking or foremilk), it may be assumed that changes in the fat content of the latter may have a certain relation to the fat content of the residual milk. This assumption is based on the fact that under certain conditions of normal milking the content of fat in residual milk may fully or almost fully pass over to the foremilk and this circumstance, obviously, cannot be reflected in the total fat content of the milk. According to our findings [188] the fat content of the residual milk of cows reached, on an average, 10.02% with fluctuations of 3.0–19.8%. In one and the same cow showing fat content of 17.4% in residual milk the fat content of foremilk was 6.8% while with fat content in residual milk at 10.6% it was only 3.7% in the foremilk.

Fluctuations in the fat content of the foremilk depend, as a rule, on changes in the fat content of residual milk. As can be seen from Fig. 4, a rise in the fat content of the foremilk to 10% coincides with high fat content in the residual milk (16.8%). However, in some experiments no connection was noticed between the fat content of residual milk and foremilk. In order to clarify this problem we subjected the data of the 990 trials to statistical examination. The correlation coefficient between the fat content of foremilk and fat content of residual milk was +0.446. This shows that there is a relation between these values in 19.8% of the experiments. The regression coefficient between the fat content of residual milk and the fat content of foremilk was 0.198. This indicates that with an increase in the fat content of residual milk by 1% the fat content of the foremilk increases by 0.2%. Construction of empirical lines of regression gave the following results (Fig. 5).

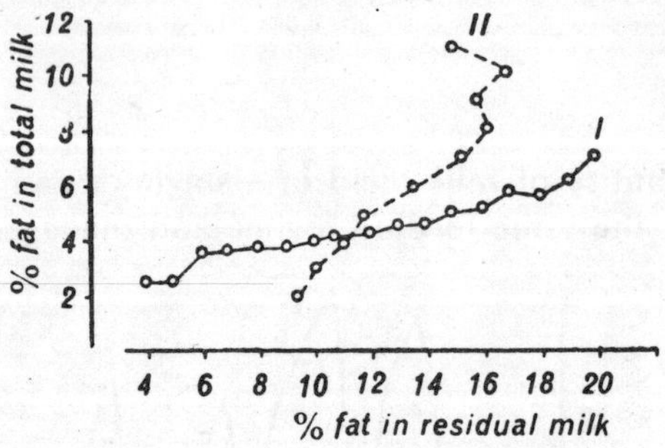

Fig. 5. Empirical lines of regression.

I—x/y; II—y/x; x—% of fat in foremilk; y—% of fat in residual milk.

A relation between the fat contents of foremilk and of residual milk is not observed in all experiments. We can see from the graph that in those experiments in which the fat content of residual milk was 4 to 10% there is no relationship between the fat content of the foremilk and that of the residual milk since the lines of regression intersect at right angles. However, when the fat content of residual milk is between 10 and 16% the lines of regression run close to one another, which points to the existence of a relation between the fat content of the foremilk and that of the residual milk. But since out of 990 trials the majority showed a fat content of 10–16% in the residual milk the assumption of such a relationship between the fat content of foremilk and that of the residual milk is quite justified. The data obtained show that residual milk may have an effect on the butterfat content of foremilk only in those cases where the fat content is high.

The changes in the fat content of residual milk are not connected with the volume of residual milk. However, Olenov [137] concluded on the basis of 15 experiments on 6 cows that there exists a correlation between the volume of residual milk and its fat content. It may be noted that the causes responsible for an increase in the yield of foremilk and in the fat content of the milk are not identical. Taking this into account, we classified the 990 trials with reference to the volume of residual milk and calculated the mean fat percentage values. The results are presented in Table 8. These data show that the percentage of fat in residual milk does not change with an increase in its volume. The correlation coëfficient between the fat content and volume of residual milk is −0.033. This shows that there is no relation between these indices. It is most likely that the presence of such a relation in Olenov's experiments is due to incomplete repeated letdown in the trial animals since with the subcutaneous doses of oxytocin used by this author not all the residual milk could have been drained off from the udder.

An increase in the fat content of milk to double the normal was observed not just in single or daily milkings. In the cow Rama it lasted over 4–5 milkings. In place of 1.0–1.5 kg of milk fat per 24-hour day, on the 23rd day of lactation she gave more than 2.4 kg and on 24th day more than 2 kg. The (fore) milk yield of the cow Rama for the 300 days of lactation was 6,175 kg with 3.7% fat content. Unfortunately physiology does not provide us with any factual material on the basis of which this phenomenon could be explained. The absence of such data in this field is also indicated by the recent work of Borsuk and Zaks [31] who observed in one experiment an increase in the fat content of milk more than twice the normal under the stimulation of pain.

Then at what cost did the milk fat increase to double the normal in our experiments? It could be on account of a direct increase in the synthesis and secretion of fat or of a withdrawal of fat synthesized and stored earlier

TABLE 8. FLUCTUATION IN BUTTERFAT CONTENT OF RESIDUAL MILK IN RELATION TO VOLUME

Number of experiments	Volume of residual milk (liters)		Percentage of fat in residual milk	
	Mean	Fluctuation	Mean	Fluctuation
222	0.36	0.1–0.5	10.12	3.0–17.4
266	0.76	0.6–1.0	11.38	6.0–19.5
165	1.29	1.1–1.5	10.71	3.8–19.8
122	1.79	1.6–2.0	11.61	6.0–19.8
59	2.31	2.1–2.5	11.21	4.8–18.9
36	2.84	2.6–3.0	11.15	4.8–18.0
28	3.31	3.1–3.5	10.39	5.0–16.5
22	3.80	3.6–4.0	11.61	7.6–17.7
24	4.41	4.1–4.5	11.04	7.6–18.6
15	4.80	4.6–5.0	10.79	5.7–18.0
15	5.33	5.1–5.5	8.11	4.8–12.3
12	5.83	5.6–6.0	9.66	5.4–17.4
4	6.28	6.1–6.5	6.28	5.1–8.7

in the interior of the secretory cells, the alveoli and lacteal vessels of the udder.

A study of the rate of secretion of fat gave the following results: In the cow Prima the rate of secretion of fat was 35–45 g/hr*. At this speed of secretion under experimental conditions, when the earlier accumulations of fat in the alveolar cavities of the udder had been eliminated, 840–1,080 g of fat was synthesized per 24-hour day, and so in 24-hour daily milkings with the milk fat content double the normal it would be possible to obtain 1,600–2,000 and even 2,456 g of fat. In view of the passive nature of the processes of synthesis of the components of milk it is difficult to assume that the usual speed of fat synthesis doubled on a particular day or two due to the effect of some factor or other.

We are left to assume that this quantity of fat had already been synthesized earlier in the udder. Its gradual withdrawal over a number of successive milkings could bring about a doubling of milk fat content. But to accept this assumption we require proof that more than the daily supply of milk fat could be stored in the cow's udder. These proofs have been obtained in our experiments. In Table 9 we give data which show that more than the daily supply of fat can be stored. By drawing off this fat the fat content of the milk in cows can be doubled.

To obtain milk with high fat content the milking should be complete. A small quantity of residual milk remains in the udder as a result of which the fat content of the foremilk is low. This testifies to the fact that it is possible to create conditions for an active letdown of milk and withdraw all the milk from the udder, including the residual milk; but in this process the foremilk will not be characterized by high fat content. Complete milking out increases the fat content of the milk only when a sufficient quantity of fat has been formed and stored in the cow's udder. An indication of this condition is high fat content in the residual milk. The same can be said about secretion. An increase in the activity of the secretory process along with partial evacuation of the udder does not double the fat in the milk. Therefore an increase in the fat content of the milk by this much is not connected directly with secretion and letdown of milk. Of course, without a high level of synthesis of milk and fat in the udder and active milk letdown it is not possible to obtain high fat content in milk in the normal milking. But these conditions are still not enough.

In order to obtain milk with high fat content a certain quantity of fat must be stored in the udder and this is bound up with the deposition of

*Rate of secretion of fat was determined thus: After two consecutive milkings the residual milk was drawn off. Then, changing the intervals, the quantity of milk and fat secreted in the udder in 1, 3, 5, 7 and 12 hours were found. Milking was induced by injection of 18 IU of oxytocin into the jugular vein.

TABLE 9. DEPOSITION OF MILK AND FAT IN THE UDDER COMPARED WITH FOREMILK AND QUANTITY OF FAT OBTAINED PER DAY

Index	Name or number or the cow					
	Prima expt. 1	Prima expt. 2	Fialka	Zorka	5364	1426
Foremilk yield in 24–hr day (liters)	20.0	23.0	5.0	11.0	21.0	25.0
% of fat	4.0	4.3	4.5	3.5	4.3	3.3
Weight of fat (g)	800	989	225	385	903	825
Single milking (liters)	7.0	7.0	1.5	4.2	11.4	4.2
% of fat	4.3	4.8	6.4	4.1	3.8	2.1
Weight of fat (g)	301	334	96	172	433	88
Residual milk (liters)	4.5	5.0	1.7	1.9	4.6	5.8
% of fat	18.6	19.8	14.4	15.6	18.0	17.4
Weight of fat	837	990	245	296	828	1,009
Total milk in udder (liters)	11.5	12.0	3.2	6.1	16.0	10.0
% of fat	9.9	11.0	10.7	7.7	7.9	10.9
Weight of fat (g)	1,138	1,324	341	468	1,264	1,097
Fat as % total fat in 24–hr day	142	134	151	121	140	134

products of secretion in the milk glands. These processes are complex, active and often accompanied by phenomena of a reversible nature. It is with just this deposition of the components of milk that the resorption phenomena are connected, a fact which was discovered in Azimov's laboratory [13]. It appears as if deposition breaks the secretory process in the milk gland, it precedes letdown of milk.

Under the term "deposition" we include the storage and transformation of secretory products in the cow's udder by means of which the fat content of the milk is raised. Thus three conditions are very necessary for achieving high fat content in the milk: (1) high level of fat secretion, (2) deposition of the synthesized fat, and (3) active letdown of milk.

It is possible to influence secretion and letdown of milk to a certain extent. But the ability of the mammary glands to deposit fat has hardly been studied. The difficulty of problems related to the study of methods of accelerating the deposition process is complicated by the fact that the act of deposition itself does not always accompany the functional state of the mammary glands. Secretion of milk may take place normally and without any deposition.

The cow's milk glands are not only influenced by the neuroendocrinal system and other internal organs but are also exposed to the influence of external factors. Dislocation of the functioning of the mammary gland induces changes in the functioning of other organs and the whole organism of the cow through the nervous and endocrinal system. Unfortunately the butter fat yield capacity of cows in this scheme has not been studied very deeply because the role of the processes of the mammary glands in respect of butterfat yield has been underestimated. The deposition of secretory products in the interior of the secretory cells is perhaps the work of the same mechanism as regulates the composition and content of the milk being secreted by the lactating animal.

Causes of retention of milk in the udder

What is the role of milk retained in the udder of the cow? On the one hand, it can be drawn off from the udder by the usual methods of milking. Therefore the residual milk cannot be considered as a constant attribute of the secretory, still less, of the synthetic processes taking place in the epithelial cells of the alveoli. Secretory processes may be accomplished with the participation of residual milk or without it. The latter, being stored in the alveolar cavities, exerts an iufluence on the nature and direction of processes proceeding in the lacteal gland. Johansson [83] believes the chief causes for retention of residual milk are the stage of lactation, interval between milkings, age and individual peculiarities of the cow. Koshi and Petersen [101] think that retention of residual milk does not depend on the stage of lactation, age of the animal or magnitude of milk yield. Our studies showed that individual peculiarities, age and level of productivity have no bearing on factors causing retention of residual milk in the cow's udder. What then are the causes of retention of milk in the udder? Dodd and Foot [49] advanced the suggestion that fluctuations in residual milk in cows depend on the effectiveness of the mode of milking. Of course, the principal causes of retention of residual milk in the cow's udder are connected with milking procedures. However, the complex of these causes includes more than the factors bound up with the milking regime. It is most likely that there are also factors connected with the maintenance and feeding regime of the cow.

The basic conclusion to be derived from the research described in this chapter may be stated as follows: The causes responsible for retention of milk in the udder are not connected with the biological peculiarities of the animal, but with the service maintenance of the lactating animals. These causes are open to concrete and thorough experimental investigation and can be corrected.

Chapter III

Physiology of Yield Increase in the Cow

It is well known that among the multitude of measures for yield increase in cows a very important place is occupied by skilled milking. The quality of machine milking is judged by the completeness of recovery of milk from the udder. On this depends the quantity of milk not only at the time but also in subsequent milkings. Therefore incomplete milking out of the udder when the cow is receiving diverse full-value fodder and conditions of maintenance have been improved and other measures connected with yield increase have been taken may lead to a lowering of the milk producing capacity of the cow. Elliott [59, 60], summarizing the literature on the effect of complete milking out on the secretion of milk, comes to the conclusion that this problem is yet to be investigated. Tverskoi [176] believes that a thorough evacuation of the milk from the alveoli during milking stimulates secretion. Injection of oxytocin in sheep [44] before milking twice a day raised milk yield by 36% as compared to control animals. Until recently the effect of incomplete milking on milk secretion as applied to machine milking in cows had not been investigated.

Effect of incomplete milking on milk yields and condition of udder

Recently Dodd and Clough [52] carried out experiments on six cows to study the effect of different degrees of incomplete milking on milk secretion and the results of this study are of great importance for a proper understanding of the essence of yield increase in cows. The experimental cows were milked twice a day by a special four-chambered milking machine with a transparent graduated cylinder. The milking regime was: vacuum 382 mm of mercury column, pulsation rate 50 per minute and pulsation ratio 1 : 1.

The yield of each quarter of the udder was ascertained over seven days. Then at each milking for 48 hours 100% of the expected milk was drawn from one of the quarters, 90% from the second, 75% from the third and 60% from the fourth. The butterfat content of the milk drawn from each quarter was determined separately. Thereafter complete milking was resumed and the milk and the fat yield of each quarter was recorded. The results of this experiment are presented in Fig. 6. It is evident from these data that incomplete milking of the quarters over two days led to reduction of the yield when complete milking was resumed immediately afterward. After resumption of complete milking the milk yield of the quarters in the first milking was more than usual because of the removal of milk that had accumulated in the quarters during incomplete milking. In the second milking the quantity of milk in each quarter diminished to the level during incomplete milking. This indicates that the extent of milking out at each milking is one factor that regulates the speed of the secretory process in the cow's udder. With incomplete milking the rate of secretion of milk rapidly declines by an amount equivalent to the quantity of milk not milked out from the quarter. Later, with complete milking over three to eleven days, the quantity of milk recovered from the quarters increased but the yield did not return to the level before the experiment. After the secretion of milk stabilized at a lower level on resumption of complete

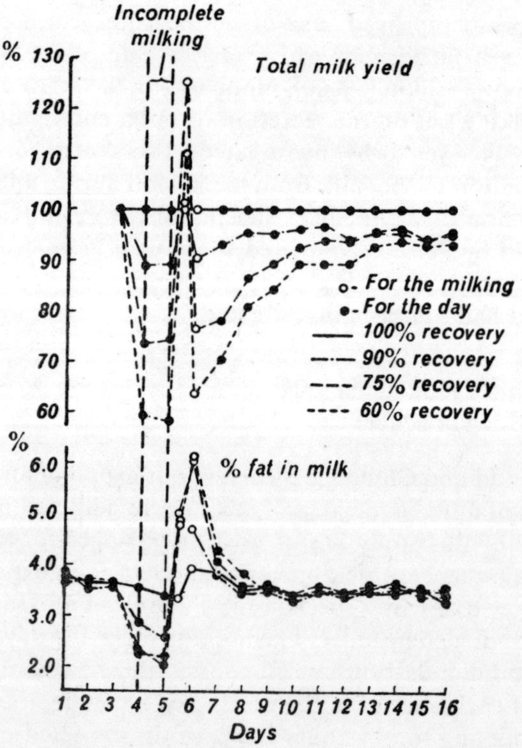

Fig. 6. Effect of incomplete milking on milk yield and butterfat content in milk (after Dodd and Clough).

milking the drop in yield was still around 5% irrespective of the degree of milking out during the experiment. The butterfat content of the milk fell in the first complete milking and then returned to normal.

Later Dodd and Clough carried out a series of experiments on the following lines. From one of the quarters the milk was drawn out completely and from the other three only 75% of the expected milk yield. Incomplete milking of one of the quarters was continued over four days, of the second over eight days and of the third over twenty days. The results of these experiments are shown in Fig. 7. As may be seen, an

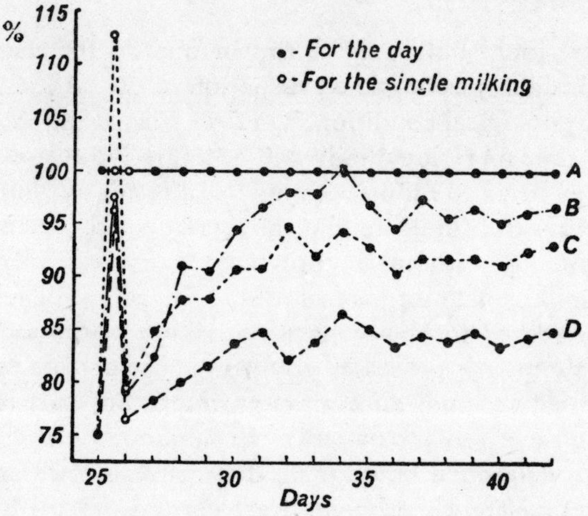

Fig. 7. Effect of partial milking on the magnitude of milk yield.

A—Complete milking of the quarters; B—Milk yield after four-day period of partial milking; C—After eight-day period; D—After 20-day period (after Dodd and Clough).

increased milk yield was observed in the first milking after restoration of complete milking out. In the next milking the yield of milk was approximately 75%. Then the yield began to climb again and this continued over a period of seven days, though without complete restoration of the former yield. Yield stabilized at a lower level. The speed and degree of recovery of milk yield after incomplete milking is inversely related to the length of the period of incomplete milking. After the four-day period of partial milking the yield diminished by 5%, after the eight-day period by 7–10% and after the 20-day period by 15–17%. In experiments conducted by Schmidt et al. [153] with machine milking, where 0.9 liter less milk was milked from the udder at each milking over a period of 10 days, the milk yield decreased in the following seven days by 6.5%. Where 1.8 liters less milk was drawn out of the udder the decrease in yield was 12.6%. In these experiments it was observed that in the period of incomplete milking there was no increase in the leucocyte content of the milk, nor were there any symptoms of mastitis.

Thus incomplete milking of the udder leads not only to a reduction in milk yield during that period but also causes a decrease in milk secretion immediately thereafter when complete milking is resumed.

Residual milk and the secretory process in the mammary gland of the cow

Since the residual milk can be drawn out by the usual methods of milking we consider the complete evacuation of the udder in milking to be the normal physiological condition. On fully evacuating the udder within a short lapse of time on residual milk will be left in the alveoli. Only after an interval do they begin to fill up with milk as a result of synthesis and secretion of milk. Our data indicate that the secretory process in the udder may take place normally without the presence of residual milk. It is obvious that residual milk, which is the result of the inhibition of letdown of milk, cannot be considered to be a constant attribute of the secretory process, still less of the synthetic processes going on in the alveolar cells.

Believing that residual milk plays an important part in the processes taking place in the alveolar cavities of the mammary gland, we ran special experiments to determine how far residual milk is necessary for normal secretion of milk. The experiments were carried out on five cows in the first half of their lactation with a milk yield of 24–29 liters per 24-hr day. The average hourly rate of milk secretion in these cows exceeded 1,000 ml/hr. Three series of experiments were run on each of the five cows—15 in all. Each series of experiments took 38 hours and involved the following: The cow was completely milked out by hand. After cessation of the milk flow in response to milking stimulation, 18 IU of oxytocin was injected into the jugular vein and the udder was milked out immediately, after which another 24 IU of oxytocin was injected into the jugular vein and milking was resumed carefully. Thus the udder was fully evacuated of all milk. If we follow Baryshnikov's reasoning [19] that residual milk includes milk that was in the secretory cells of the alveoli, then it may be assumed that in these experiments not only the alveoli but the secretory cells also were emptied of all milk.

From the moment the milking stopped after the second injection of 24 IU of oxytocin, we waited exactly 30 minutes and then injected another 18 IU of oxytocin into the cow's jugular vein and drew out the milk secreted in the interval. Then we increased the intervals to 1, 3 and 5 hours. The milking procedure followed was the same as before. Before milking

TABLE 10. FORMATION OF MILK IN THE UDDER OF COWS OVER VARIOUS EXPERIMENTAL INTERVALS BETWEEN MILKINGS

Index	Name of cow					
	Zenita	Zorka	Prima	Vesna	Krasa-vitsa	Average
Number of experiments (15)	3	3	3	3	3	3
Yield of single milking (liters)	11.4	10.5	9.8	11.9	9.3	10.6
Residual milk after 1st injection of 18 IU oxytocin (liters)	1.3	0.9	3.2	1.9	2.9	2.0
Residual milk after 2nd injection of 24 IU oxytocin (ml)	15	—	30	12	21	16
Volume of milk (ml) obtained by thorough milking with the help of injection of 18 IU oxytocin after:						
30 mins after 2nd injection	795	605	540	640	550	626
1 hour after 3rd injection	1,350	1,180	1,050	1,640	1,040	1,252
3 hrs after 4th injection	3,890	3,360	3,090	4,830	3,090	3,652
5 hrs after 5th injection	6,410	5,550	5,140	7,950	5,100	6,030
7 hrs ofter 6th injection	8,820	7,490	7,030	10,570	7,000	8,182
9 hrs after 7th injection	10,800	9,360	8,820	12,780	8,810	10,310
12 hrs after 8th injection	13,740	12,100	11,280	15,840	11,400	12,872

we injected 18 IU of oxytocin into the cow's jugular vein and drew out the milk which had been secreted over 1, 3 and 5 hours. When the intervals were increased to 7, 9 and 12 hours in consideration of the larger volume of milk stored in the udder, the milking procedure was somewhat changed.

Before injecting oxytocin the udder was emptied of the milk which by then had accumulated in the cisterns and larger lacteal vessels. After injection of oxytocin milking was resumed. This was done so that delayed milking should not inhibit the letdown of milk from the alveoli caused by the oxytocin injection. Results of these experiments are given in Table 10.

What changes does the process of milk secretion undergo after all the milk collected in the udder has been drawn out? As can be seen from the table, after complete evacuation of the udder some milk is secreted during the next 30 minutes. Hence it follows that in conditions of complete evacuation of milk from the udder the secretion of milk does not stop. But could it be that the rate of secretion of milk slows down after such a complete evacuation of the udder? According to average data the rate of secretion over the intervals of 30 and 60 minutes is the same.

These results show that in the absence of residual milk in the udder the speed of secretion does not diminish. Thus the process of milk secretion does not need the presence of residual milk and in general the presence of milk in the udder.

Speed of secretion of milk

The higher the rate of milk secretion in the intervals between milking, the higher the milk yield of the cow. What are the factors that influence the speed of secretion of milk? Studies by Tucker et al. [174] showed that prolongation of the intervals between experimental milkings from 8 to 24 hours lowered the rate of milk secretion in the udder; the fall in the speed of secretion giving a curve on the graph. These authors extracted the residual milk from the udder with the help of an intravenous injection of 20 IU oxytocin. Therefore, the quantity of milk drawn out of the udder of the animals in these experiments at different intervals accurately reflects the the speed of secretion. In our experiments, the speed of secretion was studied for intervals of 1–12 hours. The average speed of secretion at 12-hour intervals was 14.4% less than the speed of secretion at hourly intervals. These data show that when there was no milk in the alveoli of the udder the speed of secretion was higher. Gradually the storage of milk in the alveoli increased when the intervals between milkings were extended and this led to lowering of the speed of milk secretion. How quickly does the rate of milk secretion fall when the intervals between milkings are extended? In order to answer this question an estimate of the average hourly rate of secretion for various intervals was made. The estimated speed of secretion or degree of fall in this speed in proportion to the extension of the intervals between milkings was calculated this way: The rate of secretion of milk in one hour was assumed to be 100%. Subsequently, the quantity of milk formed in 1 hour was found from the quantity of milk obtained over the three-hour period. The resultant figure was divided by two and the quotient characterized the average hourly rate of secretion over the subsequent two hours. This method of calculation was followed for each interval. The indices obtained were then compared with the rate

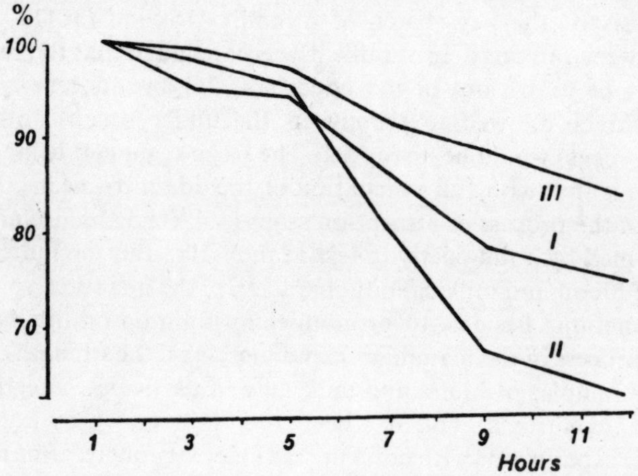

Fig. 8. Speed of secretion of milk in the udder of cows with progressively greater intervals between experimental milkings.

I—Average data; II—Cow Vesna; III—Cow Krasavitsa.

of secretion for 1 hour. The degree of fall of rate of secretion following the prolongation of intervals between milkings was calculated on a percentage basis. The results of this calculation are shown in Fig. 8. As can be seen, the secretion rate falls imperceptibly to the 5th hour. From the 5th to the 9th hour the rate falls more sharply. From the 9th hour to the 12th hour it becomes imperceptible again. The fall in the secretion rate accompanying the prolongation of intervals between milkings from 1 to 12 hours gives a curve on the graph. Thus secretion in the udder of a cow proceeds nonuniformly. These peculiarities of the secretory process in the mammary gland enable us to explain why milk yield in a cow sharply declines after incomplete milking.

Residual milk and resorption

Azimov et al. [13] reported the phenomenon of resorption in the mammary gland of the goat, i.e. reverse absorption of the products of secretion from the system of cavities into the blood. These processes have been quite satisfactorily proved to be concomitant with the secretion of milk under certain conditions [14]. However, Azimov subsequently propounded a hypothesis from which it followed that resorption is a necessary process in the mammary gland and that if it is arrested the secretion of milk is also arrested [15]. But since resorption could only involve a liquid medium, and in the mammary gland this medium is the residual milk this hypothesis made the residual milk in the udder an indispensable condition for synthesis of the new batch of foremilk [16] and fat [10] during the period between milkings. From this it was concluded that the residual milk should not be drawn out of the udder [24, 25] and in general that some milk should be allowed to remain in the udder because otherwise the secretory process would be disrupted. The basic argument behind Azimov's hypothesis is that with full evacuation of the udder by means of injections of oxytocin the process of resorption stops [13]. The absence of resorption was confirmed by a radioactivity test of the milk. But the indices of radioactivity of blood and milk cannot characterize the quantitative mechanism of milk secretion. Besides, to pronounce an opinion on the nature of the secretory process in the mammary gland merely on the strength of the radioactivity of samples of blood and milk may in particular cases lead one to erroneous conclusions. This has been very well shown by Luick [114]. In some of his experiments with cows the radioactivity of glycerine molecules in the milk fat and of the glucose were higher in the first 3 hours after milking and lower in the 10 hours after milking. This indicates that synthesis of fat and glucose proceed vigorously immediately after milking while later, 10 hours after milking, the speed of synthesis of these milk components diminishes. However, in other experiments of this series the activity of glycerine and glucose during 3 hours were, on the contrary, lower and after 10 hours higher. These data lead us to a different conclusion: that the synthesis of the components of milk is lower in the first hours after milking and higher after 10 hours. The obvious connection between these data and the completeness of letdown of milk obliged that author to recognize that this difference was caused by incomplete evacuation of the quarters of the udder before injection of the isotopes. The results of this study indicate that the investigation of the secretory process in the mammary gland is not possible without a precise reckoning of the quantity of milk formed during the experimental intervals. The resorption phenomenon in the mammary gland has been observed by other workers too [67, 110, 168], but in none of these studies do we find confirmation for the statement that resorption is a necessary condition for the secretory process in the udder of cows. This apart, Lascelles [109] showed that in sheep resorption of serous albumen with labeled iodine-131 after partial milking did not occur in the udder in the course of 4 hours while the secretion of milk continued. Azimov's hypothesis cannot easily be reconciled with the experiments by Wood et al. [201] who showed that after injection of 50 IU of oxytocin and thorough milking the synthesis of lactose and casein proceeded most intensively in the cow's udder over the next 1.6 hours, i.e. under conditions where resorption could hardly have taken place. With greater intervals between experimental milkings the synthesis of these milk components slowed down. And, finally, it is difficult to understand why the secretion of milk did not stop when the cows were milked every 2 hours [196]. It might be supposed that with the usual partial milking not all the milk in the udder had been drawn out, thanks to which resorption had taken place and this had favored secretion. But when the cows were milked every two hours with the help of injections of 10 IU of oxytocin [107, 158, 165], secretion of milk neither discontinued nor diminished compared with twice-daily milking. It is true that this dose of hormone is insufficient to drain off all the milk from the udder and hence there remained the possibility of resorption to prevent stoppage of secretion. However, in experiments [54] in which the cows were milked every hour for 156 hours with the help of internal injections of 20 IU oxytocin, milk secretion accelerated since reflex letdown of milk had been blocked [55].

With this boosted evacuation of the mammary gland there could scarcely have been requisite conditions for resorption. Yet there does not appear to have been a depressive influence on the secretory process but, on the contrary, a stimulative one. With milking of the cow every hour the rate of milk secretion increased considerably and in particular periods increased to more than double as compared to twice-daily milkings. The composition of the milk underwent no change in this period.

Considering all this it may be concluded that the first part of Azimov's hypothesis, that resorption is a concomitant of the secretory process which may under certain conditions take place in the udder, is well substantiated and need not raise any doubts. But as regards the second part of his theory that residual milk and resorption are necessary conditions for secretion of milk in the udder, we may say that all this is plainly inconsistent with the factual material gathered to date and cannot, therefore, be accepted.

At the same time, future investigation on resorption will be of exceptional interest since lactation is closely related to the water-salt metabolism of cows [32, 33], permeability of cell membranes and barrier functions of the secretory cells.

The data presented show that secretory processes in the mammary gland proceed both with and without the participation of resorption. In both cases the milk secretion in the udder is not disrupted but the direction and end results of secretion in these two cases differ. Which direction of the secretory process in the mammary gland is more advantageous for machine milking and enhancement of milk yields? At present it is difficult to answer this question. Regarding the effect of completeness of extraction of milk from the udder on the succeeding secretion, it may be presumed that complete milking with the milking machine at each milking without leaving any milk in the udder ensures a sharp increase in the productivity of milk by the cow. Therefore the elaboration of the complex of practical measures that promote complete withdrawal of milk from the udder in machine milking is a very necessary condition for an intensive increase in milk yields.

Chapter IV

Letdown of Milk in Cows with the Use of Milking Machines

Machine milking is fast replacing hand milking. But because in mastering the technique of machine milking of cows various difficulties are often met with, a section of specialists doubt the physiological parity of hand and machine milkings. Both in hand milking and in machine milking a host of arguments are advanced which, it must be regretted, are not concerned with the chief function of these methods—the complete stripping of milk from the udder. The results of our research [187] show that completeness of evacuation of milk and fat from the udder are practically the same under hand milking and machine milking.

Physiological appraisal of machine milking

Milking devices are being constantly improved and are now available in several new models of milking machines and units (installations, plants, systems). It is essential that the new milking techniques meet the special physiological features of the lactating cows. Evaluation of the physiological effects of the use of new milking machines and installations is, therefore, of considerable importance. So a method of appraisal that can satisfy scientific standards has considerable importance for uniform evaluation. The basic requirements for milking machines and installations include, apart from safety provisions control of the duration of milking and full stripping of milk from the cow's udder. To judge the completeness of stripping, it is necessary to study not only the milk and fat stripped by the milking machine but also the milk and fat retained in the udder. For this purpose, it is usual to resort to stripping by hand [5, 26, 92, 139]. The completeness of stripping the milk from the udder is determined by correlating the quantity of milk stripped by the milking machine with that stripped by hand. In the GDR, for example, a special numerical evaluation method for machine milking of cows has been worked out [46, 47]. It is based on the speed of milking and quantity of milk stripped by hand. It takes into account the skill of the milkers.

Hungarian workers, [43] however, came to the conclusion that the factors in hand milking do not provide reliable criteria for characterizing letdown in cows. This is indicated in Eisenreich's work [57] where no correlation was observed between the magnitude of hand stripping and rate of milking over the whole stretch of lactation. The method of finding the completeness of evacuation of the udder by correlating milk yield with a machine with milk yield by hand stripping was based on the assumption that hand stripping can draw out all the milk left after machine milking. But contemporary literature on the physiology of lactation contains no confirmation of this. Owing to the stoppage of the milk letdown reflex there is some residual milk, which is subject to fluctuation, left in the udder after machine milking and hand stripping. According to the data of 702 trials with 39 cows when milking machines were undergoing tests the yield of foremilk was in the range of 50–11,500 ml, hand strippings 25–4,600 ml and residual milk 50–6,500 ml. Since the quantity of milk in hand milking fluctuates widely it may be asked where the milk which is obtained by hand stripping was stored after machine milking was completed. Earlier it was thought that the milk extracted by hand stripping after machine milking is retained exclusively in the alveoli [121]. But Peinovich [141] and subsequently other workers [86, 154] showed that milk stripped by hand is located in the cavities of the teats and cisterns of the udder. However, it is difficult to believe that the 2–4 liters of milk drawn out after machine milking by hand stripping in a number of our experiments is all accommodated in the cisterns of the quarters of the udder. To check this assumption we ran a number of experiments as follows: After milking by machine was completed a plexiglass catheter was inserted into the teats of all the quarters in turn and the milk from the cisterns that had not been stripped by the machine was drained out. Then hand milking was done. Following this, an oxytocin injection was administered and the residual milk was drawn out. The results of these experiments are presented in Table 11. From these data it appears that the hand-stripped milk could have been localized in the cisternal or alveolar cavities, possibly in the minute ducts of the latter. Moreover, irrespective of its volume the hand-stripped milk could be found in both the above-mentioned cavities of the udder. The milk obtained by machine represents part of the milk stored in the udder at the start of the milking; this is milked out by the milking machine. Hand-stripped milk and residual milk comprise the unstripped part of the milk stored in the udder.

Hand-stripped milk represents the part of the milk that was let out of the alveoli during letdown into the underlying cavities but was not drawn out by the milking machine. Among the numerous potential causes of this holding back of milk may be mentioned the creeping of the teat cup on the teat toward the close of milking, weak suction of milk from the udder at the start of milking, the effect of a high vacuum on the teats, distortion of the teat base due to the high vacuum, and so on. Hand-stripped milk is an important index for evaluation of the efficiency of milking machines and installations and the study of hand-stripped milk is therefore very necessary. But the completeness of evacuation of the udder by machine milking should not be estimated on the basis of the quantity of milk obtained by hand stripping alone. After hand stripping, too, some residual milk is found in the udder which is subject to marked fluctuations. Therefore correlation of the volume of milk obtained by machine with the milking obtained by hand stripping gives incomplete, and in the majority of cases unreliable information on the completeness of milking by whatever kind of milking machine is used.

TABLE 11. DISTRIBUTION OF MILK IN THE CAVITIES OF THE UDDER OF A COW (NAME: GVOZDIK) IN MACHINE MILKING

Particulars	Number of experiment			
	1	2	3	4
Milked by the machine (ml)	9,010	7,020	7,515	10,000
Obtained through catheter (ml)	1,000	100	145	—
Hand stripping (ml)	1,400	—	80	400
Residual milk (ml)	4,600	1,600	3,210	2,000

In machine-testing stations, milking machine laboratories, construction bureaus, scientific establishments, kolkhozes and sovkhozes, the evaluation of the quality of milking machines is conducted by correlating the milk collected by the machine and by hand stripping. But now this method has lost its rationale and cannot be employed for evaluation of milking machines.

If it is assumed that hand stripping bears a relation to the completeness of the prior evacuation of the udder by the milking machine, then when large quantities of milk are obtained by hand stripping the volume of milk collected by machine should be small and when smaller quantities are obtained by hand stripping the reverse should be true. But experiments do not support this assumption. Table 12 presents the findings of 20 trials on eight cows at one and the same milking. The 20 trials are arranged in pairs such that the total of hand stripping in each pair is the same. With similar hand stripping, the completeness of evacuation of the udder by machine is found to vary. Table 13 on the other hand, concerns 16 trials conducted on seven cows in one and the same milking but with differing quantities of milk drawn by hand. The experiments were grouped in pairs on another basis whereby experiments with larger and smaller magnitudes of hand stripping were matched in each pair.

With larger yields of hand-stripped milk, owing to the fuller stripping of the residual milk the evacuation of the udder by the milking machine is higher than with smaller yields of hand-stripped milk.

Considering the above facts, we subjected the data of 702 trials to statistical analysis. The correlation coefficient between machine collection and hand stripping was found to be 0.143. In other words, there is no relation between these two factors. The correlation coefficient between machine collection of milk and residual milk is 0.259. This shows that if we exclude hand-stripped milk from the total volume of milk it is not possible to estimate the completeness of evacuation of the udder. The correlation coefficient between total milk yield (machine milk yield+hand-stripped milk yield) and residual milk is 0.9.

TABLE 12. COMPLETENESS OF EVACUATION OF THE UDDER BY MILKING MACHINE WITH SAME MAGNITUDES OF HAND STRIPPING

Indices	Experimental series	
	I	II
Number of cows	8	8
Number of trials	10	10
Machine-stripped milk (ml)	6,560	5,610
–Do– as %	82.6	77.5
Hand strippings (ml)	145	113
–Do– as %	2.0	1.4
Residual milk (ml)	1,035	1,525
–Do– as %	15.4	21.1
Total milk in udder (ml)	6,730	7,248
–Do– as %	100	100

These data show that hand stripping is inadequate as a means of evaluating milking machines. For the evaluation of milking machines and installations it is necessary to determine the completeness of evacuation of the udder on the basis of a study of milking machine yield, hand stripping yield and residual milk. But if hand stripping is omitted it must be done on the basis of milking machine yield and residual milk yield. Since residual milk is located in the alveolar spaces, the determination of the completeness of evacuation of the udder in machine milking on the basis of a study of residual milk indicates, among other things, to what extent the milking device is able to bring about a reflex letdown in the cow. According to Johansson [83] the correlation coefficient in machine milking between foremilk and residual milk was 0.61 and between foremilk and hand stripping only 0.31. Evaluation of milking machines by correlating the yield by milking machine and hand stripping lead to incorrect conclusions. Thus Korolev [100], for example, on the basis of Kaveshnikova's data that hand stripping after milking with the 2-pulse Manus machine was greater than after the 3-pulse DA–3M machine, concludes that the latter machine is better. Hence it is necessary to determine the relevance of the method of testing milking machines devised by Kaveshnikova in the Pavlov Institute of Physiology [87, 88, 89]. Evaluation of the milking machine by this method is done on the basis of the duration of the latent period of reflex letdown of milk, the quantity of milk in the aveolar part and the quantity of milk obtained by hand stripping. With the help of these indices it is not possible to determine the completeness of evacuation of milk from the udder or from the individual quarters. Therefore, the above method, in the way it is propounded, could hardly be of help for evaluation of milking machines. The following facts support this: According to Kaveshnikova [90], hand stripping after machine milking by the DA–3M gave less milk than that after using the Manus. But according to Zlaber's data [208] the magnitude of hand stripping was, on the contrary, small after using the Manus and large after using the DA–3M. In the Federal Republic of Germany, for example, 10 different makes of milking machine were tested. The magnitude of hand stripping following the use of each milking machine

TABLE 13. COMPLETENESS OF EVACUATION OF UDDER BY MILKING MACHINE WITH DIFFERENT MAGNITUDES OF HAND STRIPPING

Indices	Magnitude of hand stripping	
	Smaller	Larger
Number of cows	7	7
Number of experiments	8	8
Machine-stripped milk (ml)	5,775	5,550
–Do– as %	79.3	84.9
Hand stripping (ml)	272	515
–Do– as %	3.7	8.0
Residual milk (ml)	1,237	481
–Do– as %	17.0	7.1
Total milk in udder (ml)	7,284	6,546
–Do– as %	100	100

fluctuated in the range 0.1–0.3 kg of milk [1]. It is quite obvious that it is not possible to determine the advantage of a particular milking machine on the basis of such data.

Recently Kreilis [103] expressed the view that the quantity of residual milk obtained after oxytocin injection is not a reliable indicator for assessing the quality of milking machines. A more dependable index, according to him, was the fat content in the last drops of milk stripped from the cow's udder at the end of milking. Kreilis considers that after a thorough milking there always remains in the udder a residual quantity of milk and even with repeated intravenous injections of 10 IU oxytocin not all this residual milk can be drawn out. These views contradict Swanson and Hinton's [169] who, after a first injection of 10 IU of oxytocin, found no residual milk in the udder with further injection of 10 IU of oxytocin. This contradiction is explained by the fact that 10 IU of oxytocin is insufficient to draw out all the residual milk. Therefore the results of the experiments by these workers cannot be taken as a guide. Kreilis' work offers no experimental evidence that allows us to assume that residual milk does not have a direct bearing on the completeness of stripping of milk from the udder.

In the light of all this, we tried various methods of practical help in conducting tests (physiological evaluation) of milking machines. The basic conclusions are given below:

1. The present method of physiological evaluation of milking machines is intended for machine testing stations, milking machine laboratories, design bureaus, scientific research institutions and other organizations engaged in the construction and testing of milking machines and also for specialists in farm management working on improvement of the technology of machine milking of cows under the conditions obtaining in kolkhozes and sovkhozes.

2. The testing of milking machine must be carried out on farms free from infectious diseases where machine milking of cows has been fully mastered, and an impeccable livestock record keeping system is maintained. The plan for testing the milking machines should be interlinked with the plan of veterinary and prophylactic measures followed on the farms. Intravenous injections, etc. not required for this evaluation work are not allowed.

3. At least three milking machines are tested at the same time, and for this three groups of animals with at least seven cows in each are selected. Selection of control cows in the group is not necessary. The cows should be helped to adapt to being machine-milked and trained for this purpose. The cows should have been paired one month before testing. Animals with defective udders should not be chosen for testing. Feeding of the animals during the period of physiological testing should be adequate and consistent. Testing is done on the same farm.

4. In the physiological tests the factors considered are the time taken by machine milking, hand stripping and extraction of residual milk; volume and butterfat content of machine collected milk, hand-stripped milk and residual milk; length of intervals between milkings; and milk yield and butterfat content of milk at each normal milking during the whole period of the test. Depending on the purpose of the test these factors may be supplemented by study of the speed of milk letdown; its character as between different quarters of the udder based on a correlation of cisternal, alveolar and residual milks; duration of the latent period of the letdown reflex, and so on. Butterfat content of the milk is estimated by the acid method. Estimation of the butterfat content of the residual milk is done immediately after milking. Storage or refrigeration of samples of residual milk is not allowed.

5. To draw out the residual milk the use of synthetic oxytocin is recommended, but failing this, use can be made of Pituitrin "M" or "R" produced by national industry. Pituitrin is an extract of the posterior lobe of the hypophysis of cattle that contains the hormone oxytocin. Between 18 and 20 IU make a single dose of the hormone. Oxytocin is injected into the jugular vein. Hypodermal and intramuscular injections of the hormone are not allowed. Injections of oxytocin are given once in three days at one of the milkings. Smaller intervals between the injections are not permitted. General aseptic and antiseptic procedures are followed in giving intravenous injections of oxytocin. To disinfect the hide where it is intended to administer the injection rectified spirit may be used. A second injection should not be given at the same place on the skin or vein where a previous injection has been given. Five injections are given on the left jugular vein and the same number on the right side; they are given on alternate sides. If there is any uncertainty about oxytocin having been injected into a particular vein one injection may be skipped as a precaution. Injections should not cause any hematoma or other complications at the spot where the hide is pierced. Use of time-expired oxytocin is prohibited. The oxytocin ampoule is opened immediately before use. Irrespective of the capacity of the syringe, only one dose of the hormone at a time is drawn from the ampoule. The opening of the ampoule, drawing the hormone into the syringe and other preparatory operations should not be done in direct sunlight. For intravenous injections of oxytocin the use of a syringe of 10 ml capacity of the Rekord type and veterinary needle No. 1230 or 1022 is recommended. Use of thick-walled blood transfusion needles is prohibited. The big punctures made by such needles cause trauma to the walls of the veins, tend to bring about hematoma and intensify the pain effects, all of which make conduct of physiological test more difficult.

All the physiological work related to injections of oxytocin should be done by one particular person. A qualified milkmaid should attend to the machine milking of the cows. Besides high qualifications for this work she should possess patience, be of gentle nature and affectionate toward the animals. All these qualities should also be possessed by the cattleman, the hospital attendant, and other workers participating in the physiological evaluation of milking machines. During the physiological work shouting at the cows or treating them roughly is absolutely prohibited.

6. The cows are injected with oxytocin in the stalls. If the tests of milking machines are conducted in milking parlors with the cows untethered a special stand is erected some 5–10 m away from the milking parlor for penning the animals and administering injections. The cow should be readied for the injection after she has been stripped by the milking machine and then stripped by hand. Physiological work should not disturb the regime of milking, sequence and time of stripping prevailing in the cattle yard or milking parlor. The procedure of the treatments can briefly be put down as follows. After machine milking the cow is immediately stripped thoroughly by hand in the milking parlor or stall. Then the cow is led out to the stand or allowed to remain in the stall where 20 IU of oxytocin is injected into the jugular vein and immediately the residual milk is stripped by hand. Hand stripping of residual milk should be accompanied by a thorough massage of the udder. The duration of stripping of residual milk should not exceed 10 minutes. Special care should be taken to avoid any interruption between injection of the oxytocin and stripping the residual milk, as also between the machine milking, the subsequent hand stripping and the oxytocin injection. All these operations should follow one another without a break. The total time taken to handle the cow in this way from the beginning of the first milking to completion of hand stripping of residual milk should be regulated according to the cessation of milk flow from the teats in response to the milking action.

In the evaluation of milking machines it is necessary to observe strictly the following order of preparation of the cow to be milked: The udder is washed and its condition examined before slipping on the milking teat cups in not longer than 30 seconds. No interval is allowed between the washing of the udder and the beginning of milking by the machine. Water for washing the udder should be at about 25–30°C. Thorough massage, washing of udder with hot water and other types of stimulation by hand to bring about milk letdown are prohibited before the machine milking, which is being done for the purpose of evaluation. The milking machine should conform to the following basic requirements: it should bring about vigorous letdown of milk in cows without hand stimulation and be able to draw out the milk from the udder completely.

7. The physiological investigation of the letdown of milk should be conducted over the course of the 24-hour day cycle: in the case of morning, noon and evening milkings with three repetitions, in the case of morning

and evening milkings with two repetitions. The sequence of alternation of injections and milkings in the 24-hour day cycle is given in Table 14. Oxytocin is given to all the groups of cows on the same day. Over a period of nine days the cows are treated three times. For each group of cows milked three times a day the experiment covers 27 days, for cows with two milkings a day it covers 18 days.

TABLE 14. ALTERNATION OF OXYTOCIN INJECTIONS IN GROUPS OF COWS MILKED THREE TIMES A DAY

Group of cows	Days								
	1	2	3	4	5	6	7	8	9
1	—*	—	U	—	—	D	—	—	V
2	—	—	D	—	—	V	—	—	U
3	—	—	V	—	—	U	—	—	D

*—On these days a record is kept at each milking of the time taken for milking by machine, milk yield and butterfat content of milk.
U—Injection in the morning milking.
D—Injection in the noon milking.
V—Injection in the evening milking.

Depending on the aim of the milking machine test the physiological work on the above lines may be carried out over an extended period of time and may cover the first and second halves of the lactation period or the full period. In extended tests the oxytocin injections may be given once in 4–5 days. Changes are made in the scheme of alternation of injections in conformity with this. The 24-hour day cycle of injections of oxytocin is taken as the basis in analysis of milk letdown in cows and in the evaluation of milking devices irrespective of changes in testing schemes that may occur in response to local conditions of the farm and cattleyard and the purpose of test.

8. For comparative evaluation, each of three milking machines is tested in serial order alternately on all three groups of cows. The duration of the test period for all the milking machines for all three groups of cows should be the same. Each period should cover three 24-hour day cycles. A specimen scheme for testing milking machines is given in Table 15. The duration of the full test period is 81 days. In twice-daily milking the duration of each period would be 18 days and the number of injections 6. Total duration of test would be 54 days. With a larger number of repeated injections (4–5–6) the duration of the test for thrice-daily milking would be 108–135–162 days, with twice-daily milking 72–90–108 days. When com-

prehensive evaluation of milking machines is carried out along with physiological evaluation, they are subjected to technical, economic-zootechnical and veterinary-microbiological evaluations, all the details being coordinated on the basis of the corresponding procedure. In this process the rules and sequence of the physiological work do not change.

TABLE 15. SCHEME FOR TESTING MILKING MACHINES WITH THRICE-DAILY MILKINGS

Indices	Groups of cows					
	I		II		III	
	Days	Injections	Days	Injections	Days	Injections
Brand of milking machine	DA–3M		M–59		DPR–2	
I period	27	9	27	9	27	9
Brand of milking machine	DPR–2		DA–3M		M–59	
II period	27	9	27	9	27	9
Brand of milking machine	M–59		DPR–2		DA–3M	
III period	27	9	27	9	27	9

9. The following are the principal advantages of the new method of testing milking machines: If the rules laid down for the conduct of physiological work are observed errors in determining completeness of evacuation of milk by the milking machine are eliminated, selection of a control cow in the group is not required, there will be no necessity to conduct long tests over the full lactation period, testing of milking machines may be completed in 30–60 days, and therefore the results of the evaluation may be promptly used when new milking machines are installed. The completeness of stripping of milk from the cows will determine the quality of the machine milking. When a milking machine strips less than 70% of the milk from the udder the unsatisfactory quality of machine milking is indicated, more than 70% indicates satisfactory quality, more than 80% indicates good quality and more than 90% indicates excellent quality.

Study of milk letdown in cows in milking installations

In spite of a considerable amount of work dealing in great detail with the organization and economic aspect of machine milking, the problems of physiological analysis of the cows' reaction to the environment of the milk parlor due to the presence of various milking machines have not caught the attention of researchers. The study of all the factors that inhibit or stimulate letdown of milk in cows in milk parlors has considerable importance for elaboration of the technology of machine milking so as to permit an increase in milk yield with minimum manual labor. The use of milk parlors is economically profitable only when the productivity of cows with machine milking is not lowered by comparison with hand milking. But this is possible only under conditions where the completeness of stripping of milk from the udder by the machine in the milking parlor at each milking is the same as with hand stripping by a skilled milker. The problem then is how to organize the milking of cows in the milking parlor so as to strip the milk fully from the udder. The answer to this problem is not to be found in existing literature on the physiology of machine milking. In view of this we studied the letdown of milk in cows in milking parlors of various designs.

These studies were carried on in the sovkhoz 'Krasnaya Poima' of Lukhovitsk district in the Moscow region during an international trial of milking machines in collaboration with the Podolsk Machine-testing Station. Three milking machine installations were tested: the 'herringbone' type UDDE-16 [4], Impulse, a mobile milking parlor of the 'herringbone' type [135], and the UDM-8* with a milking stall of the walk-through type [206]. The milking stalls of the UDM-8 and UDDE-16 are placed on the same level as the yard floor, while the milking stall of the Impulse is placed a little above the level of the concrete floor. The cows are led to the raised stall by a wooden ramp.

Each milking installation worked on one herd of cows (of 150–200 head) over the whole period of the test. The cows were billeted through the 24-hour day on pasture located on the banks of the River Oka. During milking they were given feed concentrates in the milking parlors. The evaluation work began on July 24, 1961, and was completed on September 20 of the same year. Physiological testing began two months after work with the milking installations began. It was assumed that by this time the cows of each herd had become familiar with the milking installations.

For comparative physiological evaluation of the effect of milking machines and installations, a group of 13 cows was chosen from each herd. All were in the first half of their lactation. The animals were formed into groups according to types. As a result of this the average 24-hour milk yield of the cows of the experimental group in the milking installation UDDE-16 was 14.59 liters milk with 3.75% butterfat, in the Impulse 15.04 liters milk with 3.49% butterfat and in the UDM-8 15.05 liters milk with 3.79% butterfat. Two weeks after the cows had been grouped, their milk yield

*UDM-8 is a general-purpose milking parlor with eight milking stalls.

had changed from what it was at the beginning of the trial and because of this factor more than half of the animals could not be considered as reliable. The mean 24-hour daily milk yield of the cows for the test period is given in Table 16. The highest yield appeared to be from cows tested on the UDM-8, the next on the UDDE-16 and the lowest yield on the Impulse. What is this difference due to? As expected, the completeness of stripping of milk from the udder by the three milking installations seemed to be different. These data are given in Table 17. It is natural that the varying completeness of stripping affected the level of the next secretion in the udder and so the volume of milk collected. As regards the volume of milk that had accumulated in the udder at the beginning of the milking the two groups of cows milked on the UDM-8 and the Impulse were similar but due to varying completeness of stripping under the same conditions of milking the 24-hour yield of the cows milked in the milking installation Impulse was less than of those milked in the UDM-8. These results show, firstly, that the completeness of stripping of milk from the udder has a direct effect on the volume of milk yield and, secondly, that the grouping of control cows based on volume of milk yields for testing milking machines is not necessary under conditions where the completeness of stripping of milk from the udder by the milking machine is ascertained on the basis of the quantity of residual milk. It is necessary, at the same time, to ensure that the milking machines work during this period in a regular sequence on a random selection of cows.

TABLE 16. MEAN 24–HOUR DAY YIELD OF MILK OF COWS DURING TEST PERIOD

Type of milking installation	Number of cows	During the period of test		At beginning of test		At end of test	
		liters	% fat	liters	% fat	liters	% fat
UDDE–16	13	12.59	3.51	13.39	3.15	10.29	3.94
UDM–8	13	14.92	3.71	16.90	3.36	12.70	3.76
Impulse	13	11.71	2.98	13.46	3.11	10.80	2.85

Three milking machines, the DA-3M [98], DPR-2 [145] and M-59 [157], of different designs worked in a regular sequence on the group of cows in each milking installation. The use of different milking machines for testing milking installations is necessary for the following reasons: Milking installations of diverse design are equipped with milking devices which differ in the regime of work. Therefore, differences in the reaction of the cows observed in each milking parlor or installation were governed by

TABLE 17. CHARACTERISTICS OF MILK LETDOWN IN COWS DURING TEST PERIOD

Factors	Type of installation		
	UDM–8	Impulse	UDDE–16
Number of cows*	13	13	13
Number of experiments	234	234	234
Duration af milking (min sec)	5.33	5.02	3.48
Single milking yield (liters)	5.62	4.20	4.14
Hand stripping yield (liters)	0.36	0.48	0.31
Residual milk (liters)	0.99	2.46	1.12
Total milk in udder (liters)	6.97	7.12	5.57
Speed of milking (ml/min)	1,030	834	1,088
Machine-stripped milk (%)	80.6	58.8	74.3
Hand-stripped milk (%)	5.2	6.7	5.5
Residual milk (%)	14.2	34.5	20.2
% fat in foremilk	4.20	3.19	3.89
% fat in hand stripping	9.90	6.40	7.51
% fat in residual milk	13.19	10.99	9.87
% fat in total milk of udder	5.78	6.08	5.30
Fat extracted by machine (%)	58.6	30.8	54.6
Fat extracted by hand stripping (%)	8.9	7.0	7.8
Fat extracted from residual milk (%)	32.5	62.2	37.6

*In the following tables the results of tests of milking machines have been given for groups of 13 head of cattle for each machine.

two factors: the milking installation and the milking machine. To find out which of these two components causes inhibition or stimulation of milk letdown in the cows in each milking installation, the milking machine with its own pulsator and bucket was transferred from one installation to another for milking the test group of animals. In this way it was possible to know the differences in the reaction of the cows to the milking machines and installations and the related environmental factors.

Testing was done with thrice-daily milking. As far as possible fixed intervals were maintained between milkings. The duration of the intervals between milkings was precisely fixed. During the test period the length of interval between the milkings was on an average up to 10 hrs 43 min for morning milking, 7 hrs 54 min for noon milking, 6 hrs 49 min for evening milking. Milk letdown was studied according to the following indices: duration of milking of the cow by the milking machine, quantity of milk withdrawn by the machine (machine collection), butterfat content of this milk, duration of hand stripping, quantity of milk withdrawn in hand stripping, butterfat content of this milk, time taken to draw out the residual milk, quantity and butterfat content of residual milk.

Letdown of milk in cows was studied in each installation for definite periods at all three daily milkings—morning, noon and evening—and the data obtained were analyzed. Average data obtained from a few 24-hour cycles were taken as a basis for the analysis of letdown. The scheme followed for testing the installations is given in Table 18. In the basic period of testing the letdown was studied in two series of experiments: (a) without massage of the udder during milking, (b) with massage of udder for 1 min 30 sec after slipping the teat cups on the teats.

TABLE 18. SCHEME OF TESTING MILKING MACHINES AND INSTALLATIONS

Particulars	Type of milking installation					
	UDM–8		Impulse		UDDE–16	
	Days	No. of trials on each cow	Days	No. of trials on each cow	Days	No. of trials on each cow
			Basic period			
Milking machine	DA–3M		M–59		DPR–2	
Experiments without massage	18	6	18	6	18	6
Experiments with massage	18	6	18	6	18	6
			Closing period			
Milking machine	DPR–2		DA–3M		M–59	
Experiments without massage	9	3	9	3	9	3
Milking machine	M–59		DPR–2		DA–3M	
Experiments without massage	9	3	9	3	9	3

The study of the role of massage in milking plants is governed by the following circumstances. It is known that massage of the udder stimulates letdown in machine milking of cows but its effectiveness and expediency are not clear.

Besides this, there were grounds to believe that where massage has a very great effect on completeness of udder evacuation, it is the design of the milking system that creates a more favorable environment for the onset of milk letdown. In the entire period of tests control milking was conducted once in 5–10 days and always on the day before the test. In this test also the duration of stripping of milk and the fat content of the milk in each batch of milk collected were studied. These control milkings were done with the milking machine used in the milking system on the day of test.

Letdown of milk was studied like this: Immediately after stripping of milk by the milking machine thorough hand stripping followed in the milking parlor stall. The milk was collected in a milk pail. Next the cow was led out of the parlor into the special pen and secured, and an injection of 18 IU of oxytocin was given in the jugular vein. The residual milk was stripped by hand right away. The total time taken to handle one animal from the start of machine milking to the conclusion of stripping of residual milk averaged 21 min 58 sec. The injection of oxytocin was given at one milking only every third day.

To ascertain the completeness of udder evacuation by machine milking the quantity of milk obtained by the machine, the hand strippings and the residual milk were totaled and then the separate components were expressed as percentages of this total. The yield of butterfat was also calculated in this way. As a base for calculation of milk ejection the yield of one milking in the 24-hour daily cycle was taken. The mean indices of the letdown calculated for one milking were worked out from equal numbers of experiments run at the morning, noon and evening milkings. The data obtained in these experiments served as a basis for physiological evaluation of the effect of using milking machines and installations.

Effect of massage of udder on milk ejection with machine milking

Massage has a positive influence on the development of the udders of cows and heifers [45, 129]. Several researchers [93] recommend massaging the udder for a minute or two at the start of milking, halfway through and at the close of milking in order to increase milk yield. Udder massage is hard manual work. The basic aim of the improvement of the machine milking technology of cows is to minimize the use of the hand on the udder. Therefore the practice of using a large amount of manual labor in machine milking has hardly any future. Syusyukin in the case of cows [170] and Grachev [71] in the case of goats showed that the maximum withdrawal of milk from the alveolar cavities is observed where the duration of the massage of the udder before milking coincides with the latent period of the letdown reflex, i.e. when it is done within about 30–60 seconds. If milking is done three minutes after the letdown reflex there is a decrease in the foremilk yield of 19.6% and in butterfat content of 0.65%. The above authors did not ascertain the residual milk and this makes it difficult to evaluate their findings. In view of this we studied the effect of the duration of udder massage before milking on the completeness of letdown in cows. The experiments were run in the Lenin kolkhoz in Krasnodar circle of Labinskii district on five cows at the morning milking 10 hours after the last milking. First the latent period of the letdown reflex, i.e. the period from the start of stimulation of the udder nerve endings to the start of letdown, was ascertained. The cows were hand milked into a four-chambered milk pail, the rear quarters being stripped first and then the front ones. No supplementary operations such as washing and wiping of the udder were carried out. In all five series of experiments were run with 25 trials on five cows in each series. In the first series, no massage was done prior to milking, in the second the udder was massaged for 20 seconds, in the third the duration of the massage coincided with the latent period of the letdown reflex (33–65 seconds), in the fourth the massage was done for 70 seconds and in the fifth for 120 seconds. Immediately after massage the rear quarters were milked. After the front quarters were milked an injection of 18 IU oxytocin was given in the jugular vein and residual milk stripped from all the quarters. The oxytocin injection was administered once in 24 hours at the morning milking. Results of the 125 trials are presented in Fig. 9. From these data we see that udder massage before milking inhibits letdown of milk. In view of this, to study the effect of udder massage on completeness of stripping of milk by machine the udder was not massaged before milking in subsequent trials, as is usually recommended, but was massaged after slipping on the teat cups over the teats. The details of milk collection in milking installations without udder massage are given in Table 19, and with massage in Table 20. From these data we see that mass-

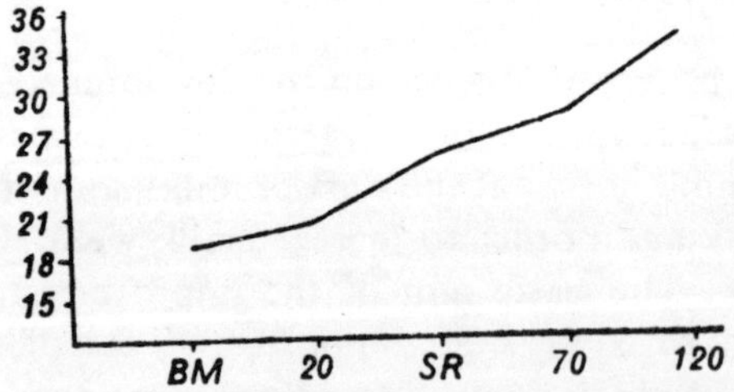

Fig. 9. Changes in the volume of residual milk in right rear quarter in relation to duration of udder massage before milking.

BM—without massage, SR—udder massage during latent reflex period of letdown (in seconds), cow Angara—43, Zenita—37, Grusha—51, Mongolka—65, Ordinara—33. Vertical axis—*residual milk as percentage of total volume of milk of quarters.* Horizontal axis—*duration of udder massage in seconds.*

age of the udder after slipping on the teat cups over the teats stimulates the letdown of milk. The volume of milk and fat thus collected in the Impulse is lower by 20% than in the UDM-8. Though in both the UDM-8 and the Impulse the same amount of milk accumulates in the udder before each milking due to incomplete stripping of milk from the udder in the Impulse its mean single milking yield was 1.9 liters less than the yield obtained in the UDM-8, the decrease being 33%.

Thus incomplete stripping and increase in residual milk in the udder lead to a marked lowering of milk productivity. The volume of hand strippings is fairly close in all three types of milking installations. Based on these data the erroneous conclusion may be drawn that the evacuation of milk by these milking installations is of the same order.

TABLE 19. LETDOWN OF MILK IN COWS WITHOUT MASSAGE OF UDDER

Particulars	Type of milking installation		
	UDM–8	Impulse	UDDE–16
Milking machine	DA–3M	M–59	DPR–2
Number of trials	78	78	78
Duration of milking (min sec)	7.15	5.42	3.37
Yield of single milking (liters)	5.75	3.85	4.06
Butterfat (%)	3.15	2.63	3.44
Speed of milking (ml/min)	793	675	1,121
Milk withdrawn by milking machine (%)	74.70	54.30	69.39
Butterfat collection by machine (%)	44.65	24.79	45.08
Hand stripping (liters)	0.70	0.55	0.55
–Do– as %	9.10	7.75	9.35
Residual milk (liters)	1.25	2.69	1.24
–Do– as %	16.20	37.95	21.26
Total milk collected from the udder (liters)	7.70	7.09	5.85

The analysis of residual milk allows us to avoid this error when evaluating milking machines.

The volume of hand strippings after udder massage decreases in all three types of milking installations and the greatest decrease is observed with the UDM-8. The quantity of residual milk also diminishes in the UDM-8. As a result of udder massage in this installation hand strippings decrease by 450 ml or 64%, and residual milk by 400 ml or 32%. After

TABLE 20. EFFECT OF MASSAGE OF UDDER ON LETDOWN OF MILK IN COWS

Particulars	Type of milking installation		
	UDM–8	Impulse	UDDE–16
Milking machine	DA–3M	M–59	DPR–2
Number of trials	78	78	78
Duration of milking (min sec)	6.02	5.07	3.31
Yield of single milking (liters)	5.97	4.32	4.20
Butterfat (%)	4.82	3.50	4.11
Speed of milking (ml/min)	990	844	1,193
Milk collected by milking machine (%)	84.50	59.70	74.84
Butterfat collected by milking machine (%)	66.00	33.50	56.50
Hand strippings (liters)	0.25	0.35	0.27
–Do– as %	3.50	4.90	4.86
Residual milk (liters)	0.85	2.55	1.14
–Do– as %	12.00	35.40	20.30
Total milk collected from the udder (liters)	7.07	7.22	5.62

massage, the completeness of udder evacuation in the UDM-8 rather reflects two components: 5.6% is at the expense of hand-stripped milk and 4.2% at the expense of residual milk. The same pattern is also observed in respect of fat content. An increase in the fat content of 21.4% takes place, this being at the expense of the fat content of hand strippings (10.6%) and of residual milk (10.8%). These data have great significance for the interpretation of the physiological effect of udder massage. The decrease in hand-stripped milk is explained by the fact that massage, by stimulating the relaxation of the smooth muscles of the udder cisterns and teat sphincters, facilitates the withdrawal of milk from these cavities by the milking machine. Because of this the quantity of milk left for hand stripping diminishes.

The phenomenon of the decrease in residual milk and fat after massage is significant in another way. It is known [184] that udder massage causes an outflow of milk from the alveolar cavities of the udder. Zaks [204] maintains that milk ejection from the alveoli cannot be brought about with the help of massage, but his view is not confirmed by experiments. Residual milk may not be found in the cisterns, it may be distributed in the alveolar cavities of the udder. Hence a decrease in the volume of residual milk after massage is conclusive proof that massaging at the start of milking before slipping on the teat cups stimulates milk ejection from the alveoli. From this it is evident that massage of the udder can bring about reflex ejection of milk.

A diminution in hand strippings after machine milking by 4.5% and in residual milk by 5.2% was also observed by Rostovshchikova [150], who employed the technique of washing the udder with warm water at 50–52°C before milking and massaging it in the 5th minute of milking. The insignificant decrease in hand strippings is apparently due to the fact that massage of the udder in the 5th minute of milking has little effect on milk letdown. As far as the volume of residual milk is concerned, these data require further refining. The hypodermic injections of 5 to 6 ml pituitrin employed in the above experiment are inadequate to effect withdrawal of milk.

Now let us look at the effect of massage on completeness of stripping of milk from the udder in machine milking in relation to the duration of intervals between milkings. It is known that with longer intervals between milkings the cow's udder contains more milk. This intensifies the ejection of milk and increases the speed of milking. It is assumed that the greater the accumulation of milk in the udder before milking, the more complete will be the evacuation of milk from the udder. Marchenko [117], for example, in his study of variations in residual milk in the cow at different intervals came to the conclusion that on average its volume was constant but as a percentage of the total volume of milk in the udder, i.e. when fully stripped, it varied. As acknowledged by the author himself, this consistency is observed only if the conditions of milking and the volume of 24-hour milk yield are maintained at a relatively constant level [119]. But such uniformity in the conditions of milking and in the maintenance regime of cows in cattle yards is the exception rather than the rule. Hence Marchenko's data must be considered as only a small part of the total picture of milk letdown in cows. There is much more to be done before uniformity of conditions and maintenance regime become the normal technology of milking cows in dairy establishments.

Results of the study of milk letdown in cows in the UDM-8 in respect of length of intervals between milkings is given in Tables 21 and 22.

With the UDM-8 8.89 liters of milk accumulates in the udder before the morning milking, 8.20 liters before the noon milking and 6.04 liters before the evening milking. These data show that owing to the variability of the milk letdown of cows in milking installations a similar amount of milk may accumulate over different intervals between milkings and conversely, different amounts of milk may accumulate over similar intervals.

Massaging the udder stimulates the letdown of milk and fat in cows at all three milkings irrespective of the length of the intervals between milk-

TABLE 21. MILK EJECTION WITH UDM–8 WITHOUT MASSAGE OF UDDER IN RELATION TO LENGTH OF INTERVAL BETWEEN MILKINGS (USING MILKING MACHINE DA–3M)

Particulars	Milkings		
	Morning	Noon	Evening
Number of trials	26	26	26
Interval between milkings (hrs min)	10.26	7.50	7.29
Duration of milking (min sec)	8.36	7.47	5.23
Speed of milking (ml/min)	784	795	803
Single milking yield (liters)	6.75	6.19	4.33
–Do– as %	76.00	75.58	71.66
Hand stripping (liters)	0.48	0.98	0.66
–Do– as %	5.34	11.90	10.94
Residual milk (liters)	1.66	1.03	1.05
–Do– as %	18.66	12.52	17.40
Fat % of milk collected by machine	3.03	3.22	3.22
Fat % of hand-stripped milk	9.93	11.07	6.80
Fat % of residual milk	13.31	13.37	11.00
Fat in milk collected by milking machine (%)	43.24	44.92	46.48
Fat in hand-stripped milk (%)	10.00	24.28	14.97
Fat in residual milk (%)	46.76	30.80	38.55

ings. As a result of massaging the udder the butterfat content of the milk stripped by the milking machine sharply increases by 1.1% in the morning, 2.14% at noon and 2.09% in the evening.

Massaging the udder brings about a change in the accumulation of milk in the udder according to the length of interval between milkings. In these trials 8.87 liters milk had accumulated in the udder at the morning milking, 6.75 liters at noon and 5.62 liters in the evening. If the intervals between milking were increased there was an increase in the quantity of milk stored in the cow's udder before milking. But this had no effect on the letdown of milk. These data show that the completeness of milk and fat collection with the UDM-8 does not depend on the length of the intervals between milkings or the total volume of milk accumulating in the udder before milking. At first sight these results seem to be in stark contradiction with Marchenko's results. The data on the completeness of evacuation of the udder in Marchenko's experiments could not reflect

TABLE 22. EFFECT OF MASSAGE OF UDDER ON LETDOWN OF MILK WITH UDM–8 IN RELATION TO LENGTH OF INTERVALS BETWEEN MILKINGS (USING MILKING MACHINE DA–3M)

Particulars	Milkings		
	Morning	Noon	Evening
Number of trials	26	26	26
Interval between milkings (hrs min)	10.44	7.51	6.40
Duration of milking (min sec)	5.46	6.25	5.57
Speed of milking (ml/min)	1,493	869	800
Single milking yield (litres)	7.60	5.57	4.77
–Do– as %	85.73	82.57	85.01
Hand strippings (liters)	0.22	0.31	0.22
–Do– as %	2.43	4.60	3.84
Residual milk (liters)	1.05	0.87	0.63
–Do– as %	11.84	12.83	11.15
Fat % of machine collected milk	4.13	5.36	5.31
Fat % of hand stripped milk	10.47	10.59	9.83
Fat % of residual milk	14.63	13.45	15.73
Fat collected by milking machine (%)	64.05	66.66	67.00
Fat in hand-stripped milk (%)	4.59	7.34	6.58
Fat in residual milk (%)	31.36	26.00	26.42

incomplete repeated letdown of milk since the residual milk of the animals he used was drawn out with the help of intravenous injections of 18 IU oxytocin. The methodical side of his experiments was, as we see, strictly maintained. The discrepancy in milk letdown as between Marchenko's and our own experiments must be due to the influence of diverse external factors. Milking installations create a new setting for letdown which is very unlike the conditions for milking the animals in barns. The study of the effect of these conditions on the character and completeness of letdown of milk in machine milking is of great practical importance.

It may be asked whether we can ascertain the completeness of stripping of milk from the udder in machine milking on the basis of the butterfat content of the next lot of milk obtained in response to massage and hand stripping. From the data in Tables 21 and 22 it is evident that the volume of strippings from the udder is greater at the morning milking than at noon, but the butterfat content of hand-stripped milk, on the other hand, is lower at the morning milking than at noon. By employing massage the volume of milk obtained is appreciably increased but this is not reflected in changes in the butterfat content of hand strippings. Similar results have been obtained at other milking plants. In order to determine the completeness of stripping by the milking machine on the basis of the indices of butterfat content of hand strippings it is necessary to work out transfer coefficients which would indicate a mathematically precise relation between changes in the butterfat content of hand strippings and the thoroughness of milking. But the possibility of obtaining such transfer coefficients seems to be ruled out since there is no relation between the butterfat content in hand stripping and the volume of milk obtained with the milking machine. At the same time we may emphasize that determining the butterfat content in hand-stripped milk and in residual milk is a necessary factor in the evaluation of milking machines. These indices are used in the calculation of the completeness of fat ejection.

With the UDDE-16 milking installation (Table 23) massaging the udder had a greater effect on the completeness of milking by the milking machine at the morning milking. The increase in the volume of milk obtained after udder massage is at the expense of hand strippings and residual milk. An increase in fat ejection also takes place at the same time at the expense of a decrease in the fat content of hand-drawn and residual milk. At the noon milking, the effect of massaging the udder is less. The effect of udder massage on completeness of milk collection at these two milkings is less with the UDDE-16 than with the UDM-8. The results of massaging at the evening milking are quite unexpected. The volume of milk stripped by the milking machine at the evening milking after udder massage was less than without massage. These data show that in conditions where reflex milk ejection is partly inhibited, the massaging of the udder cannot cancel out the effect of inhibition and so the use of massaging in milking installations in similar conditions seems to be ineffective.

The absence of any reaction to massage at the evening milking may be explained by the short interval since the noon milking, during which less milk accumulates than before the other milkings.

Before the morning milking there was an accumulation of 6.30 liters of milk, before the noon milking 5.34 liters and before the evening milking 5.21 liters.

With the Impulse milking installation (Table 24) massage had no effect on the completeness of evacuation of the udder by the milking machine at the morning milking when the largest quantity of milk accumulated in the udder. When external factors come into operation, causing inhibition of milk letdown, massage does not have any stimulating effect on milk letdown in the cow irrespective of the degree of fullness of the udder or duration of the interval between milkings.

TABLE 23. EFFECT OF MASSAGE OF UDDER ON COMPLETENESS OF MILK LETDOWN IN COWS IN UDDE–16 INSTALLATION IN RELATION TO LENGTH OF INTERVALS BETWEEN MILKINGS (USING MILKING MACHINE DPR–2

Particulars	Milkings		
	Morning	Noon	Evening
	Without massage		
Number of trials	26	26	26
Intervals between milkings (hrs min)	10.42	8.00	6.52
Milk withdrawn by machine (%)	74.62	61.10	71.91
Hand strippings (%)	7.82	11.88	8.48
Residual milk (%)	17.56	27.02	19.61
Fat obtained by machine (%)	49.16	38.92	47.44
Fat in hand strippings (%)	13.96	15.62	11.42
Fat in residual milk (%)	36.88	45.46	41.14
	With massage		
Intervals between milkings (hrs min)	10.26	7.43	6.46
Milk withdrawn by machine (%)	82.06	72.94	68.07
Hand strippings (%)	4.36	5.14	5.18
Residual milk (%)	13.58	21.92	26.75
Fat obtained by machine (%)	62.08	56.71	49.76
Fat in hand strippings (%)	8.29	7.04	5.45
Fat in residual milk (%)	29.63	36.23	44.79

With the Impulse, milk letdown without udder massage is small at all three milkings and particularly at the noon and evening milkings. Use of massage at the morning milking did not improve letdown but, on the contrary, changed it for the worse.

Massaging the udder stimulates milk letdown in cows only under conditions where the milking plant creates favorable circumstances for reflex letdown. Whenever milk letdown is disturbed completely or even slightly massaging the udder exerts little or no effect on the letdown of milk. There may be instances where massaging the udder in milking installations even reduces the letdown of milk in the animal. Massaging the udder during machine milking is a laborious business. Hence it is

TABLE 24. EFFECT OF MASSAGE OF UDDER ON COMPLETENESS OF MILK LETDOWN IN COWS IN IMPULSE INSTALLATION IN RELATION TO LENGTH OF INTERVALS BETWEEN MILKINGS (USING MILKING MACHINE M–59)

Particulars	Milkings		
	Morning	Noon	Evening
	Without massage		
Number of trials	26	26	26
Intervals between milkings (hrs min)	10.28	7.57	6.36
Milk withdrawn by machine (%)	65.97	43.28	49.99
Hand strippings (%)	8.67	5.07	9.31
Residual milk (%)	25.36	51.65	40.70
Fat obtained by machine (%)	31.74	18.37	24.10
Fat in hand strippings (%)	12.36	3.24	9.92
Fat in residual milk (%)	55.90	78.39	65.98
	With massage		
Intervals between milkings (hrs min)	10.10	8.03	6.26
Milk withdrawn by machine (%)	58.00	53.97	67.56
Hand strippings (%)	5.88	3.68	4.98
Residual milk (%)	36.12	42.35	27.46
Fat obtained by machine (%)	29.65	29.20	42.49
Fat in hand strippings (%)	6.32	2.37	5.96
Fat in residual milk (%)	64.03	68.43	51.55

desirable to devise procedures that would ensure high values of stripping of milk from the udder by the milking machine without massaging.

Milk letdown in cows in the UDM-8 installation

According to the findings of the basic test period, when there was no massaging, the completeness of milk letdown in cows milked in the UDM-8 was higher than in the UDDE-16 by 5.31% and higher than in the Impulse by 20.40%. As we have already seen, the external conditions for triggering the milk letdown reflex are more favorable with the UDM-8 than with either the UDDE-16 or the Impulse. But the UDM-8 is equipped with the three-phase machine DA-3M. Would it be possible to improve the letdown indices in this installation with the help of other milking machines? As can be seen from Table 25, the completeness of stripping of milk and fat is similar with the milking machines DPR-2 and DA-3M. The completeness of stripping of milk by the M-59 is higher as compared with the DA-3M and DPR-2 by 6.17%. Completeness of fat ejection is higher by 15.74% as compared with the DA–3M and by 13.39% as compared with the DPR-2. These results were obtained in one and the same milking parlor and on the same cows under conditions where no procedure or treatments was employed on the udder that could have stimulated letdown of milk. Compared with the DA-3M, the increase in the completeness of milking by the milking machine M-59 was achieved at the expense of the hand strippings, down 4.92% and residual milk, down 1.26%, and compared with the DPR-2 by 2.45% and 3.69% respectively. The quantity of milk by hand stripping after milking with the M-59 was lower as compared with the DA-3M by 450 ml or 64%; as compared with the DPR-2, by 210 ml or 46%. The quantity of residual milk compared with the DA-3M was down by 350 ml or 29% and as compared with the DPR-2 by 370 ml or 30%. It is apparent from these data that the M-59, which works on the two-track principle with paired teat cups, causes a highly active reflex milk letdown while working at high speeds and as a result more milk is withdrawn from the alveoli of the udder. But the DPR-2, though it has as high a speed of milking as the M-59, can withdraw only as much as the DA-3M does at low milking speeds.

The indices of completeness of stripping of milk from the udder by means of the M-59 have a great practical importance in view of the following circumstances: There was no massage or other hand stimulation of the udder to promote milk letdown before slipping the teat cups of the M-59 over the teats. In spite of this the M-59 caused an active letdown of milk: the completeness of letdown was higher by 4.38% in comparison with skilled manual stripping. The volume of fat ejection was higher by 3.59%. On average, hand stripping removed from the udder 76.5% of the milk and 56.8% of the fat [187]. These facts show that in the UDM-8 installation certain conditions are present that allow the M-59, even without manual stimulation, to strip a larger quantity of milk and butterfat than with hand stripping. Consequently with the use of the UDM-8 there is the posssibility of replacing all hand labor by machines for stripping milk from the udder. It follows from this that with machine milking of cows in milking installations there is a real possibility of increasing the milk yields of the animals without handling the udder since increasing the milk yields of cows with machine milking and raising their productivity depend directly on complete stripping of milk from the udder at each milking.

Significant results have been obtained in the analysis of the volume of

TABLE 25. MILK LETDOWN IN COWS IN UDM-8 INSTALLATION

Particulars	Brand of milking machine		
	DA-3M	M-59	DPR-2
Number of trials	78	39	39
Duration of milking (min sec)	7.15	3.48	4.05
Single milking yield (liters)	5.75	4.88	5.11
Single butterfat (%)	3.15	3.83	3.59
Speed of milking (ml/min)	793	1,285	1,252
Milk withdrawn by machine (%)	74.70	80.88	74.74
Fat obtained by machine (%)	44.65	60.39	47.00
Hand strippings (liters)	0.70	0.25	0.46
-Do- as %	9.10	4.18	6.63
Residual milk (liters)	1.25	0.90	1.27
-Do- as %	16.20	14.94	18.63
Total milk in udder (liters)	7.70	6.03	6.84

stripping from the udder by the M-59 operating in the UDM-8 according to intervals between milkings. The results of this study are presented in Table 26. From these data it can be seen that with the replacement of the DA-3M by the M-59 the volume of milk ejection at the morning milking with the same group of cows rose by 7.2% and fat ejection by 14.0% at the expense of hand stripping, which decreased by 0.23 liter, and residual milk, down by 0.62 liter. The increase in the volume of strippings of milk and fat by the M-59 compared with the DA-3M at the noon milking was greater than at the morning milking. This increase was also at the expense of the volume of hand strippings and residual milk, which diminished by 0.77 liter and 0.39 liter respectively. At the evening milking the volume of stripping by the M-59 is the same as with the DA-3M but fat withdrawal by the M-59 is higher by 7.72% than by the DA-3M. These data show that installed in the UDM-8 the M-59 working at much higher speeds draws out from the udder, without hand stimulation of letdown, a considerably larger quantity of milk and fat in a much shorter time than the DA-3M. It can be seen from Table 25 that the DA-3M, with udder massage, draws out a larger quantity of milk at the morning milking and a larger quantity of fat at the evening milking. But the M-59 is able to extract from the udder a similar quantity of milk and fat at the noon milking without the udder being massaged.

The data presented so far offer a somewhat different approach to the appraisal of the role of massage of the udder in machine milking of cows.

Massage stimulates milk letdown by milking machines; however, in our experiments, a higher value for completeness of withdrawal of milk and fat from the udder was not obtained after massaging in comparison with machine milking unaccompanied by massage. All this taken together shows that manual massage of the udder in machine milking cannot be considered as a factor that must be employed in dairy installations whether now or in the near future, for raising extraction of milk and fat from the udder of cows and also increasing the speed of milking. The search for factors that intensify milk ejection in cows without the use of manual massage or any other kind of manual stimulation is of great practical importance.

TABLE 26. COMPLETENESS OF MILK AND FAT LETDOWN USING THE MACHINE M–59 IN MILKING STALL UDM–8 DEPENDING ON THE INTERVAL BETWEEN TWO MILKINGS

Particulars	Milkings		
	Morning	Noon	Evening
Number of trials	13	13	13
Intervals between milkings (hrs min)	11.29	8.04	6.35
Duration of milking (min sec)	4.27	3.33	3.25
Speed of milking (ml/min)	1,438	1,394	964
Single milking yield (liters)	6.40	4.95	3.29
–Do– as %	83.20	85.30	71.20
Hand stripping (liters)	0.25	0.21	0.30
–Do– as %	3.80	3.70	6.80
Residual milk (liters)	1.04	0.64	1.03
–Do– as %	13.00	11.00	22.00
Butterfat obtained by milking machine (%)	57.20	69.30	54.20
Butterfat from hand strippings (%)	5.50	6.00	7.10
Butterfat from residual milk (%)	37.30	24.70	38.70

Data on milk letdown obtained with the UDM-8 indicate that the design of this installation comprises external features relatively favorable for the onset of letdown in cows. This installation has separated milking stalls of the 'walk-through' type situated side by side. The stalls are separated by smooth partition walls. This design of milking stall protects the cows from needless agitation at the time of milking. The animals enter the stand from the pre-milking platform. On this platform they wait quietly for their turn while other cows are milked in the stalls. This normal tranquillity at milking and relaxation before milking are necessary elements in the preparation of the cow. As a result of this the cows enter the milking stall well prepared. Milk letdown in response to the milking action sets in rapidly and the udder is more completely emptied of milk. The milking stands of the UDM-8 are designed in such a way that the cows feel at ease. There is no risk of traumatic injuries to the skin or bruises or wounds being caused by projecting parts of the milking stall, partition walls and troughs. All these ensure the satisfactory preparation of the animals to be milked in the UDM-8. It is well known that among the numerous factors favoring successful preparation of the cow for milking an important place belongs to conditioned reflex ejection of milk. On the basis of the factor of completeness of letdown obtained with the UDM-8 we may presume that the design of the milking plant produces a positive conditioned reflex reaction in the cows which intensifies milk letdown at milking time.

It may, however, be mentioned that the cow's genial state at milking time, and this includes condition reflexes intensifies milk letdown in the installation only where the unconditional component or direct milking influence are reasonably effective. This explains the fact that massage of the udder intensifies milk letdown just as substitution of the DA-3M by the M-59 does. Therefore future research on improvement of the technology of milking cows in milking installations should concern itself with improvement of the planning and design of milking stands as well as with the refinement of the milking machine.

In spite of the high mean values of volume of milk letdown with the UDM-8 in some series of experiments, higher than analogous values with hand milking, as a whole the milk letdown of cows in this installation appears to be subject to sharp fluctuations, though the high values for completeness of withdrawal of milk obtained with the UDM-8 show that this installation creates favorable external conditions for the onset of milk letdown.

Letdown of milk in cows in milking installation of the UDDE-16 type

The UDDE-16 installation is intended for simultaneous milking of eight cows. It has two milking sections with eight stalls on each side. The cows are brought to the milking section in groups of eight and stand in a 'herringbone' pattern at an angle of approximately 30° to the drain running down the middle. In recent years our national industry has been producing modernized milking installations of the 'herringbone' UDE-16B type [207].

Particulars of the milk letdown of cows using the UDDE-16 are given in Table 27.

These data show that in the UDDE-16 installation the M-59 and DA-3M withdraw from the udder more milk and fat than the DPR-2. The work indices of the M-59 and DPR-2 (speed of milking, completeness of stripping milk and fat) when installed in the UDDE-16 are inferior to the indices in the UDM-8 whereas the work indices of the DA-3M are somewhat better, though this improvement pertains only to the completeness of fat ejection and speed of milking. The volume of hand stripping and residual milk as percentages of total milk in the udder with the M-59 and DPR-2 are larger in the UDDE-16 than in the UDM-8. This shows that in the UDDE-16 these machines perform worse not only in withdrawing milk from the udder cisterns but also, owing to partial inhibition of milk letdown, withdraw the milk less completely from the alveolar cavities of the udder. These data also indicate that partial inhibition of letdown in the UDDE-16 is the effect of the installation and the complex of surroundings it creates.

TABLE 27. MILK EJECTION IN COWS IN UDDE–16 INSTALLATION

Particulars	Brand of milking machine			Average
	DPR–2	M–59	DA–3M	
Number of trials	78	39	39	156
Duration of milking (min sec)	3.37	3.50	4.39	4.02
Speed of milking (ml/min)	1,121	1,139	811	1,009
Milk collected by machine (liters)	4.06	4.36	3.77	4.07
–Do– as %	69.30	77.00	75.10	73.70
Hand strippings (liters)	0.55	0.22	0.26	0.34
–Do– as %	9.40	3.90	5.20	6.20
Residual milk (liters)	1.24	1.08	0.99	1.11
–Do– as %	21.30	19.10	19.70	20.10
Butterfat % in machine milked collection	3.44	3.45	4.14	3.66
Butterfat % of hand strippings	7.85	6.78	6.29	7.22
Butterfat % of residual milk	10.24	9.90	9.65	9.95
Fat withdrawn by machine (%)	45.00	55.20	58.30	52.50
Fat content of hand strippings (%)	13.80	5.40	6.10	8.70
Fat content of residual milk (%)	41.20	39.40	35.60	38.80

Results with the DA-3M when working in the installation UDDE-16 were somewhat different from those in the UDM-8. Hand strippings as a percentage with the DA-3M in the UDDE-16 were lower than in the UDM-8. Residual milk as a percentage was, on the contrary, greater in the UDDE-16 than in the UDM-8. In the same way, all the milking machines withdrew less milk from the alveolar cavities of the udder with the UDDE-16 than with the UDM-8. Similar results were obtained in respect of completeness of fat extraction. All these data indicate that letdown of milk in cows in the UDDE-16 is not as good as in the UDM-8.

The partial inhibition of milk letdown in cows in the UDDE-16 is also confirmed by the fact that in the UDM-8 udder massage hastens letdown and permits high indices of completeness of milking (82–85%) to be reached at all three milkings, whereas in the UDDE-16 this high value (82%) for completeness of milking with udder massage is obtained only at the morning milking. At the noon and evening milkings such values for completeness of stripping were not achieved.

What could be the causes of this inhibition of milk letdown in the UDDE-16? It could perhaps he presumed that in the separate stalls of the UDM-8 the conditions for the onset of milk letdown were more favorable than in group milking sections in the 'herringbone' pattern. In group milking sections the possibility of excitement of the cows at milking time cannot be ruled out, and this may cause inhibitive influences on the reflex of milk letdown. It is obvious that this problem calls for a thorough experi-

TABLE 28. COMPLETENESS OF WITHDRAWAL OF MILK AND FAT BY THE M–59 IN RELATION TO LENGTH OF MILKING INTERVALS IN THE UDDE–16 INSTALLATION

Particulars	Milkings		
	Morning	Noon	Evening
Number of trials	13	13	13
Intervals between milkings (hrs min)	10.42	8.19	7.26
Speed of milking (ml/min)	1,359	1,038	922
Milk stripped by machine (%)	84.1	68.1	77.2
Hand stripping (%)	3.0	5.5	3.2
Residual milk (%)	12.9	26.3	19.6
Fat extracted by machine (%)	62.1	44.3	59.3
Fat in hand stripping (%)	5.3	7.0	4.5
Fat in residual milk (%)	32.6	48.7	36.2

mental investigation in view of its importance for the practical aspect of machine milking of cows.

The working characteristics of the M-59 milking machine in the UDDE-16 installation are given in Table 28.

When installed in the UDM-8 the M-59 milks out without udder massage 83–85% milk at the morning and noon milkings, but with the 'herringbone' arrangement it milks out this much milk only at the morning milking. In the case of the UDDE-16 this high performance at the noon and evening milkings is not matched by the M-59 along with udder massage. The indices of completeness of stripping milk and fat by the DA-3M working in the UDDE-16 are given in Table 29.

TABLE 29. COMPLETENESS OF STRIPPING MILK AND FAT BY THE MACHINE DA–3M IN RELATION TO DURATION OF INTER-MILKING INTERVALS IN THE INSTALLATION UDDE–16

Particulars	Milkings		
	Morning	Noon	Evening
Number of trials	13	13	13
Intervals between milkings (hrs min)	11.08	7.58	6.49
Speed of milking (ml/min)	1,042	724	640
Milk withdrawn by machine (%)	86.1	64.1	73.3
Hand stripping (%)	3.5	7.0	5.1
Residual milk (%)	10.4	28.9	21.7
Fat extracted by machine (%)	70.2	40.4	64.0
Fat in hand strippings (%)	6.3	6.3	5.4
Fat in residual milk (%)	23.5	53.3	30.6

In the UDDE-16 both the DA-3M and the M-59 give high value for completeness of stripping milk and fat only at the morning milking. It is worth mentioning here that the highest mean values in this series of experiments for completeness of milk and fat ejection in cows were obtained at the morning milking in the UDDE-16 using the DA-3M without udder massage. These results show, firstly, that for the DA-3M there are conditions in which it can withdraw without massage a somewhat larger quantity of milk and fat than the M-59 and, secondly, that these favorable conditions for milk letdown in cows may be created in the UDDE-16 type installation also. In spite of the fact that on an average the data concerning the values of milk letdown in the UDDE-16 are inferior to those in the

UDM-8, the values for completeness of stripping milk and fat by the DA-3M at the morning milking without udder massage in the UDDE-16 points to the possibility of effective stripping of milk in a 'herringbone' installation by milking machines without manual stimulation of the udder. These data enable us to come to another conclusion of great practical importance in the adoption of machine milking of cows. In group milking of cows it is possible to obtain in individual series of trials high values for letdown of milk without manual stimulation of the udder.

In some isolated milkings it is possible to obtain very high values of milk ejection in herringbone-type installations. As can be seen from Table 30, in the UDDE-16 we find conditions in which the milking machines milk out by far the larger part of the milk and butterfat in some two-three minutes while the share of hand stripping and residual milk amounts to only 6–10%. These data show that in the herringbone arrangement there exists great potential for organization of effective machine stripping of milk in cows without manual stimulation of the udder. Complete milking of cows with such high effectiveness would permit increased milking of milch animals in dairy establishments with minimum use of manual labor. We may recall here that full stripping of milk at each milking is the basis of enhanced milk yield in cows and so no high-production machine milking technology will be effective in the final analysis if the values of complete

TABLE 30. EFFECT OF SPEED OF MILKING ON COMPLETENESS OF UDDER EVACUATION IN COWS (UDDE–16)

Particulars	Number of cow					
	9020		8662		8822	
Date of experiments (in 1961)	Sep. 1	Aug. 26	Aug. 2	Sep. 11	Aug. 26	Aug. 8
Milking machine	M–59	DPR–2	DPR–2	DA–3M	DPR–2	DPR–2
Duration of milking (min sec)	3.0	5.0	2.0	3.5	3.0	5.0
Milk stripped by machine (liters)	5.0	4.6	4.0	4.4	7.0	4.8
Milk stripped by machine (%)	93.1	86.1	94.0	88.1	90.0	82.0
Milk stripped by hand (liters)	0.17	0.30	0.07	0.12	0.13	0.05
Residual milk (liters)	0.2	0.45	0.2	0.48	0.6	1.0
Residual milk (%)	3.7	8.4	4.6	9.6	7.7	17.1

stripping of the udder remain low. Unfortunately, however, in the practice of machine milking and particularly in the introduction of newer dairy installations in kolkhozes and sovkhozes sufficient attention is not paid to these problems. Consequently the productivity of cows often diminishes in dairy plants and this, in turn, leads to the discrediting of machine milking of animals on dairy farms.

In the UDDE-16 the milking machines milk out the cow in 4 minutes on an average; the rate of milking exceeds 1 liter/min. A high speed of milking in the herringbone arrangement is attained owing to the organization of milking, which conforms to the physiological regularity of the cows' milk letdown reflex. Two milkmaids work in the 'herringbone'. Each attends to eight cows on one side of the installation. Four minutes are spent in washing the udders and slipping on the teat cups (at 30 sec per cow). In other words, as the teat cups are being slipped onto the teats of the eighth cow the first is completing her milk letdown. Thus the cycle of work of the milkmaid on either side of the 'herringbone' will be continuous. This order of work excludes any possibility of interruption of operations. Washing and wiping of the udder, slipping on the teat cups and examination of the state of the udder after milking—all these activities are carried on one after the other without a break. All these factors allow maximum utilization of the milk letdown reflex activity which appears in the first minute after the start of milking [189].

Milk ejection in cows in the Impulse installation

The results of our study of milk ejection in cows in the Impulse parlor are given in Table 31.

From these data it may be seen that the speed of milking with the M-59 and DPR-2 in the Impulse type installation is lower but with the DA-3M it is higher than in installations of the UDM-8 and UDDE-16 types. In the Impulse the milking machines milk out altogether just a little more than half of the milk. The rest of the milk is retained in the udder. The volume of hand strippings and residual milk increases. With the Impulse the larger part of the butterfat is retained in the residual milk.

In the UDM-8 installation equipped with the M-59 collection of milk is higher by 26.58% and that of fat higher by 35.60% than in the Impulse installation. From these data we see that the milk letdown reflex in the Impulse is sharply inhibited, as a result of which the milking machines generally draw out of the udder only that part of the milk that has accumulated in the cisterns of the teats and quarters and in the larger lacteal vessels, the so-called 'cisternal milk,' which usually comprises about one-half of the total quantity of milk in the udder. To draw out this milk it is only necessary to overcome the resistance offered by the teat sphincters.

Under the influence of the vacuum the teat sphincters open and the milking device draws out from the cistern the milk that has accumulated there in the interval between milkings. Owing to inhibition of letdown in the Impulse the milking machine is not in a position to draw out the alveolar milk. It is this, for example, that is responsible for the low butterfat value of milk obtained by the milking machine installed in the Impulse. Milk withdrawn by the M-59 has an average butterfat content of 2.63% and that obtained by the DPR-2 only 2.51. On the other hand as a result of increased milk withdrawal, the milk drawn by the DA-3M has a butterfat content of 3.31% which in comparison to the M-59 is 8.93% higher.

TABLE 31. LETDOWN OF MILK IN COWS IN THE IMPULSE

Particulars	Make of milking machine		
	M–59	DPR–2	DA–3M
Number of trials	78	39	39
Duration of milking (min sec)	5.42	4.14	4.58
Rate of milking (ml/min)	793	884	933
Volume of milk withdrawn by machine (liters)	3.85	3.74	4.63
–Do– as %	54.30	55.93	63.23
Hand strippings (liters)	0.55	0.48	0.76
–Do– as %	7.80	7.17	10.37
Residual milk (liters)	2.69	2.46	1.93
–Do– as %	37.90	36.90	26.40
Fat extracted by machine (%)	24.79	23.31	35.04
Fat in hand stripping (%)	8.57	7.34	11.97
Fat in residual milk (%)	66.64	69.35	52.99

From the data given in Table 31 it may be seen that an inferior quality of milking is found with all milking machines in the Impulse installation. This points to the fact that the low values for full withdrawal of milk and fat in the Impulse are not related to the performance of the milking machines and least of all to the machine M-59. The Impulse installations are equipped with milking machines of the M-59 type. Since the cows seem to be habituated to the M-59 in the Impulse installation it is naturally expected that the completeness of withdrawal of milk and fat from the udder by the M-59 will be higher in comparison with the DPR-2 and DA-3M. However this was not borne out. The M-59 gave approximately the same results as the DPR-2 in the Impulse. But installed in other types of plant the M-59 showed better results than the DPR-2 and the DA-3M.

All this shows that inhibition of milk letdown in cows is the effect of the Impulse milking installation and the environment it creates.

The values for extent of withdrawal of milk and fat by the DPR-2 in the Impulse installation are given in Table 32. The low values obtained with this machine in this installation are seen at all three milkings. We notice that in this installation, notwithstanding the high speed of milking at the morning and noon milkings the DPR-2 withdraws from the udder in all little more than half of the milk and approximately one-fourth of the butter-fat component. In the UDM-8, withdrawal of milk by the DPR-2 at a milking speed of 1,486 ml/min is 15% higher than in the Impulse at the morning milking. This indicates that because of inhibition of milk letdown the milking machine is able to draw out only the cisternal milk, which contains less fat. The fat content of milk obtained by the DPR-2 in the Impulse installation reached 2.15% at the morning milkings, 3.01% at noon and 2.59% in the evening.

Fat content of the total milk accumulated in the udder reached 5.18% in the morning, 6.63% at noon, and 6.73% in the evening. As we see from these data, the alveolar milk rich in fat is not withdrawn from the udder by the milking machine in the Impulse installation owing to inhibition of milk letdown. In Table 33 we find that in the Impulse installation the

TABLE 32. EXTENT OF WITHDRAWAL OF MILK AND FAT BY MILKING MACHINE DPR–2 IN RELATION TO INTER-MILKING DURATIONS IN IMPULSE INSTALLATION

Particulars	Milkings		
	Morning	Noon	Evening
Number of trials	13	13	13
Intervals between milkings (hrs min)	12.39	7.25	7.11
Milking speed (ml/min)	1,204	1,050	412
Milk withdrawn by machine (liters)	5.62	3.78	1.82
–Do– as %	66.0	57.0	38.0
Hand strippings (liters)	0.44	0.76	0.24
–Do– as %	5.0	11.1	5.2
Residual milk (liters)	2.50	2.13	2.77
–Do– as %	29.0	31.9	56.8
Fat extracted by machine (%)	27.0	26.0	14.2
Fat in hand stripping (%)	8.1	9.2	4.0
Fat in residual milk (%)	64.9	64.8	81.8

DA-3M gave like the DPR-2, low indices for extent of milk withdrawal and butterfat extraction at all three milkings.

The material presented so far shows that in the Impulse installation various types of milking machines gave similar low values for extraction of milk and fat from the udder. In none of the trial series were satisfactory values obtained in this installation for completeness of milking by milking machines. Nor did massage of the udder show any stimulative effect on milk letdown in this installation. The values for withdrawal of milk and fat remained low as before. All these considered together show that when milk letdown in cows occurs in unfavorable surroundings direct milking action is not effective. As a consequence of this the letdown reflex is partially or completely inhibited and the milking machine will not be in a position to withdraw alveolar milk.

TABLE 33. EXTENT OF WITHDRAWAL OF MILK AND FAT BY MILKING MACHINE DA–3M IN RELATION TO INTER-MILKING DURATIONS IN IMPULSE INSTALLATION

Particulars	Milkings		
	Morning	Noon	Evening
Number of trials	13	13	13
Intervals between milkings (hrs min)	10.27	8.03	7.31
Milking speed (ml/min)	1,053	911	829
Milk withdrawn by machine (liters)	5.09	4.92	3.88
–Do– as %	66.10	67.10	55.13
Hand strippings (liters)	0.76	0.94	0.58
–Do– as %	10.0	13.0	8.12
Residual milk (liters)	1.81	1.43	2.56
–Do– as %	23.9	19.90	36.75
Fat extracted by machine (%)	38.1	40.1	27.0
Fat in hand strippings (%)	14.0	15.2	9.0
Fat in residual milk (%)	47.9	44.7	64.0

Unfavorable surroundings for the onset of the milk letdown reflex in cows reflecting design peculiarities in the Impulse type of installation result in a situation where inhibition of milk letdown is very frequently observed, at one milking after another. During the period of physiological tests, out of 234 trial milkings in each type of installation inhibition of letdown of milk was observed in the UDM-8 in a total of 19 cases, in the UDDE-16 in 33 cases, but in the Impulse in 101 cases. This also illustrates the unsatisfactory quality of machine milking in installations of the Impulse type.

To what extent does continuous undermilking affect the animal's milk productivity? The two groups of cows, one of which was milked in the Impulse installation and the other in the UDM-8 were of approximately equal milk production potential. The animals of both herds were under practically the same feeding and maintenance regimes. However, due to the fact that the quality of milking in the Impulse installation was not as good as in the UDM-8, the milk yield in the former turned out to be lower, though the animals possessed equal physiological ability. Before physiological testing the average quantity of milk that accumulated in the udder before each milking was found to be similar in cows of both groups (in the Impulse and UDM-8 installations). But the average single milking yield of the animals in the Impulse in the test period was lower by 1.42 liters than in the UDM-8. On an average with thrice-daily milking each of the 13 experimental animals in the Impulse showed 4.26 liters less yield than the cows in the UDM-8 per 24-hour period. The decrease in output for 30 days for each experimental cow amounted to 127.8 liters. According to the data of economic research the decrease in output in the Impulse installation reaches 200 liters per cow per month. Of course, with this loss in milk productivity economy in labor cannot compensate the cost in production and as a consequence the use of the milking installation becomes economically unprofitable.

TABLE 34. INFLUENCE OF MILKING INSTALLATION ON MILK PRODUCTIVITY OF COWS

Type of installation	Number of cows	Milk yield per cow per month in liters		
		June	July	August
Impulse	154	275	237	195
UDM-8	146	475	437	383
Difference		200	200	188

How can we explain this low quality of machine milking in the Impulse installation and why does this installation set up such unfavorable conditions for the onset of the letdown reflex in cows? First of all we may look at the rigidity of construction of the Impulse milking installation. The milking section is constructed of welded metal tubes which have a zigzag look. The inner walls of the milking stalls in both sections consist of alternating short and long panels at right angles to one another. This design solution for milking stalls on the 'herringbone' pattern is convenient, for instance for transportation of the milking installation, but in practice, it is incapable of providing a congenial atmosphere for cows being milked. It has been noted with the Impulse that the rigid, crammed design of the milking sections has led to the maximum number of bruises and traumatic injuries to cows on the tuber coxae, the legs, root of the tail and other projecting parts of the body as the cows enter the milking stalls. As a result of this a protective reaction is produced in the animals in relation to the installation. When they enter the milking installation the cows first of all endeavor to position themselves in such a way that the projecting parts of their body do not come into contact with any jutting corner in the partition walls. Very often, afraid of being bruised, the cows assume unnatural poses such as stooping, bent back and hind-quarters tucked in, neck stretched too far ahead with the body also straining forward, and so on. Under these conditions it is difficult, of course, to imagine that a positive reaction to the milking installation can be generated in the cow prior to milking. On the contrary, there arises in the animal a negative reaction to the entire complex of surroundings associated with the Impulse milking installation. Consequently on entering the milking stall the letdown reflex is inhibited, the milking stimulation of the udder becomes ineffective, and this leads in the final analysis to complete or partial inhibition of milk letdown and detention of a large quantity of residual milk in the udder.

Special consideration has to be given to the cows' approach route to the milking stalls. The floor of the milking section of the Impulse type of installation is raised 75 cm above the concrete floor. The milking plant is fixed on 20 jacks (10 on each side) in working condition on the concrete. The cows enter and leave the milking stall along a wooden ramp with guard rails. The frame is made of wooden planks so that the hooves of the animals may not slip as they walk. How did the cows react to this kind of arrangement? First of all, in negotiating this ramp the tips of the hoofs—the so-called third phalanx—often broke as they struck the boardwalk planks. Often the fore or hind legs would slip between the wooden planks, leading to loss of balance. Our observations showed that as a result of this a large number of cows showed an investigation reflex in regard to the wooden ramp, sniffing it and studying its condition before stepping on. A number of them usually tried to jump across it, especially when leaving the installation. This led to unnecessary vibration of the milking sections, and tended to produce noises that disturbed the animals inside and led to inhibition of milk letdown. Therefore this arrangement for entry of cows into the milking section of an installation is hardly to be recommended for adoption by the dairy industry.

The yard from which the animals enter the milking stalls should be at the same level as the floor of the installation. The floor should be even, the side walls or partitions of the entrance zone should have no sharp angles and as far as possible should be made of pliant but simple material. The milking stalls of the Impulse installation are of the open type. Therefore all kinds of distracting factors that could inhibit the letdown of milk exert their influence on the cow at milking time. We may also note the behavior of the cows in the enclosure before milking and in the milking stalls. In the corral near the installation of the UDM-8 type the cows stand quietly or lie down before milking. Some chew the cud. In this state the cows, without being urged, spontaneously approach the milking stall. But in the case of the Impulse installation the animals are first driven into a corral and then into a small pen near the milking installation. From here the cows approach or are driven into the milking stall. When the animals are driven into the pen their placid state is dispelled, they stop chewing the cud and some try to escape. It was in this state that cows were usually found in the milking stalls in the Impulse installation. It is only to be expected that all these things, causing inhibition of milk letdown in cows, lead ultimately to unsatisfactory performance in machine milking in the Impulse type of installation.

We have to keep in view the fact that quite often the physiological evaluation of milking installations does not coincide with the technical evaluation. The problem then is how to come to a decision. Milking systems which inhibit letdown of milk by the animals should not be recommended to the dairy industry. The technical evaluation of dairy plants consists as a rule in testing the various parts and working mechanisms for reliability and stability of work regime, wear and tear, breakage and other such factors. It is on this basis that the characteristics of the technical performance of the milking installation are defined. The principal problem of physiological rating of this or that milking parlor is to indicate what qualities of technical performance the system lacks and to what extent these were lacking when the system was used with living subjects—the cows. Therefore the physiological rating of milking installations should have a decisive voice in the recommendation of milking installations to the dairy industry. The Impulse installation, for example, includes a number of technical novelties which without any doubt are of interest for the development of more improved milking installations. Firstly we may mention that the Impulse is equipped with a thoroughly up-to-date system with milking machines of the M–59 type working in pairs, which in the UDM–8 and UDDE–16 installations have surpassed the DA–3M and DPR–2 in working indices. Impulse installations have a quite original vacuum system with a central pulsator, a milk system with milking devices, a milk conducting system and vacuumized milk tank, a pulsating mechanism for washing and disinfection of the milking systems and other features fully described by Novikov [135]. However, neither the design nor the method of construction of the milking stalls in the Impulse is quite satisfactory. The latter feature leads to the creation of unfavorable conditions for the onset of the milk letdown reflex and consequent lowering of productivity of the animals.

Testing of milking machines is usually done by machine testing stations, all of which together function at an all-Union level under the name "Sel'khoztekhnika." In testing milking machines the technical appraisal of new types of milking techniques takes pride of place owing to the fact that this type of activity requires above all engineering-cum-technical specialists. In addition to technical appraisal, at best a microbiological study of the work regime of the milking machines is made. The completeness of stripping of milk from the udder is tested on the basis of comparison of machine collection and hand stripping. It was by this method that the Impulse and UDM–8 milking installations were evaluated in 1961 at the Podolskii Machine-testing Station. The Impulse installation was deemed the better one. All this shows that there is an imperative need to review the existing system of government testing of milking machines.

Milk letdown in cows in rotary-type installation KDU–1

It is well known that milking installations of the rotary type have the capacity to extract high productivity from the work of milkmaids. A few such rotary milking installations have been introduced in Italy [105] and other countries. In each, some 1,000–1,500 cattle are milked. The mastering of the technology of rotary milking installations requires first of all the solution of a number of problems of a biological nature: the investigation of the reflex reactions of the cows to the moving platform, the estimation of completeness of milking, the elaboration of techniques for fighting mastitis, and so on. In the Donetz agricultural station, 400 cows were switched to the rotary installation KDU–1 from hand milking during the period December 1964–June 1965. This provided an opportunity to study milk letdown in cows milked by this system. The results of this study are given in Table 35. Initially milk letdown was studied over the 24–hour cycle, first at the morning milking and then at noon and in the evening too. Injection of 20 IU oxytocin was given in the jugular vein once in four days. The extent of milk letdown was estimated in hand milking before the cows were transferred to the KDU–1, and then again after the cows had been transferred and got accustomed to it.

In the KDU–1 the letdown of milk was studied in the following way: After the animal had been milked by the milking machine she was hand stripped right there on the platform. Then after the cow had left the platform she was tethered, the oxytocin injection was administered and the residual milk stripped.

TABLE 35. INFLUENCE OF ROTARY PARLOR KGU–1 ON MILK LETDOWN IN COWS

Particulars	Mode of milking	
	Manual	Machine
Number of cows	8	8
Number of trials	24	24
Single milking yield (liters)	4.31	3.37
–Do– as %	85.3	72.2
Hand stripping (liters)	—	0.51
–Do– as %	—	10.8
Residual milk (liters)	0.74	0.81
–Do– as %	14.7	17.0
% butterfat in milk yield	3.81	4.18
% butterfat in hand stripping	—	8.54
% butterfat in residual milk	9.36	11.60
Fat in milk harvest (g)	168.5	141.1
–Do– as %	65.6	50.5
Butterfat in hand stripping (g)	—	43.78
–Do– as %	—	15.6
Butterfat in residual milk (g)	68.93	94.12
–Do– as %	34.4	33.9

The milking machines in the KDU–1 are of the two-track type, re-engineered on the basis of the triple device DA–3M. In the KDU–1 there are 21 separate milking stands in the ring arranged in tandem or one behind the other [146]. The procedure is as follows: As the cow approaches the milking stand and receives her portion of fodder the milkmaid drops the teat cups into a pail containing water at 50–55°C to warm them up, then washes the udder, wipes it with a towel and slips the teat cups onto the teats. These operations take some 20–30 seconds. After the teat cups have been slipped on the teats the udder is massaged for 1.0–1.5 minutes. Hand stripping of cow after machine milking in the KDU–1 was not generally done but eight cows were hand stripped at the noon milking to find out the completeness of withdrawal of milk from the udder. The duration of milking by the machine in the KDU–1 did not exceed four to five minutes, the full turn of the circle being completed in six to seven minutes.

The completeness of milking by the machines in the KDU–1 was less than by hand milking by 13.1% in milk and 15.1% in fat. The quantity of milk withdrawn by the milking machines in the KDU–1 together with that obtained by hand stripping comes to about 83% of the total milk in the udder, i.e. almost as much milk as is withdrawn in hand milking. Milk withdrawn in hand stripping is the part of the milk which mostly comes out of the alveoli and underlying cavities of the udder but is not stripped by the milking machine. The quantity of residual milk both in milking by hand and in the KDU–1 is almost the same. These data indicate that the KDU–1 may bring about reflex milk letdown of a high order in cows.

Cows react differently to a milking installation of the KDU–1 type. Table 36 shows that of the eight experimental cows the completeness of milking in the KDU–1 diminished by comparison with hand milking in all but two cows. The volume of residual milk in the udder of cows milked in the KDU–1 was less when they were hand milked.

This means that the letdown reflex in response to the milking action in cows milked in the KDU–1 was higher than when they were hand milked, and in consequence of this the alveolar spaces of the udder were more completely emptied of their milk.

TABLE 36. COMPLETENESS OF LETDOWN OF MILK IN COWS IN RELATION TO THE MODE OF MILKING

Serial No. of cow	Hand milking				Machine milking					
	Milk collection		Residual milk		Milk collection		Hand stripping		Residual milk	
	liters	%	liters	%	liters	%	liters	%	liters	%
10980	3.4	86.1	0.54	13.9	2.8	70.8	0.70	17.6	0.46	11.6
11933	6.0	91.6	0.51	8.4	4.7	71.6	0.58	8.3	1.28	19.1
13721	3.0	87.9	0.41	12.1	2.2	69.2	0.41	13.1	0.55	17.7
11943	4.7	86.8	0.71	13.2	4.5	81.5	0.36	6.7	0.64	11.8
12903	4.2	85.0	0.73	15.0	3.3	83.1	0.28	6.2	0.48	10.7
11940	4.7	76.8	1.41	23.2	3.6	67.7	0.80	15.0	1.25	17.3
11925	3.7	80.2	0.90	19.8	2.4	70.6	0.31	9.7	0.66	19.7
13759	4.9	87.6	0.68	12.4	3.6	66.4	0.66	12.3	1.15	21.6

The retention of milk which was subsequently stripped by hand is not due to inhibition of the letdown reflex by external factors. It is brought about by the unsatisfactory work of the functional working parts of the KDU–1 and first of all the milking machines. The design of these machines in the KDU–1 is not satisfactory. It is not possible to attain a regime of constant vacuum and normal work of pulsators, nor of the work of collectors and the rubber of the teat cups. As a result a considerable quantity of milk is detained in the udder that has already entered the underlying cavities from the alveoli owing to the letdown reflex. But the milking machine is unable to withdraw it owing to erratic working. This is then stripped by hand. It is essentially on account of this factor that there occurs a deterioration in the values for completeness of withdrawal of milk and fat from the udder when the cows are milked in the KDU–1. These data also indicate that milking the animals on a rotating platform in individual stalls of the rotary installation KDU–1 is not antagonistic to the natural letdown reflex of the animals. In a rotary milking installation it is possible to bring about milk letdown as effectively as in other types of installation or even in stalls in cowsheds. It may also be mentioned here that there are many factors in rotary installations that may cause a full or partial inhibition of letdown of milk. The impact of these factors on letdown in the rotary type of milking installations has not been researched. This omission should be repaired since there is as yet no biologically well founded technology of milking cows in rotary installations.

The results of the research on milk letdown in cows in the KDU–1 show that taking the cows onto a moving platform is not an unnatural act. The cow's reaction to the movement of the platform, the sensitivity of the cow's vestibular apparatus to movement of the platform, and above all the degree and nature of the influence of these reactions on the completeness of withdrawal of milk and fat by the milking machine are some of the more important problems deserving study in the course of adoption of rotary installations in milk production.

Discussion of the results of physiological testing of milking installations

The data obtained in the physiological testing of milking installations indicate that the quality of the installation has to be assessed chiefly by the indices of the letdown of milk. But apart from physiological tests milking installations are subjected to technical, economical and veterinary-cum-biological evaluations. It is also important that the installation should provide certain conveniences for the operating personnel. The results of all these evaluations of milking installations should be utilized not only in estimating their profitability for industry but also for studying possibilities of further improvement in design. A discussion on some of these possibilities is of considerable practical interest.

The extent of letdown of milk and fat is the highest in the UDM–8 installation lower in the UDDE–16 and lowest in the Impulse. The quality of milking in the UDM–8 is good, in the UDDE–16 and KDU–1 satisfactory and in the Impulse unsatisfactory. The lowest rate of milking is found in the Impulse.

In some cases the quality of machine milking is estimated on the basis of the volume of single and 24-hour milk yields. Based on this standard it may be argued that the quality of machine milking in the Impulse and the UDDE–16 is the same since the magnitude of single and 24-hour day milk yields in the test periods is similar. But this conclusion would not be justified because the similarity of the mean single milking yields of cows in the Impulse and the UDDE–16 arises from a different cause. The small milk yield of cows in the UDDE–16 as compared with the UDM–8 is explained by the fact that the potential productive capacity of the experimental animals in the UDDE–16 is lower than in the UDM–8. The total volume of milk which accumulated in the udder before milking was 1.4 liters less in the UDDE–16 than in the UDM–8, and due to this the single milking yield appeared to be lower by 1.48 liters.

The volume of the single milking yield of the cows in the Impulse installation can be accounted for by other cause. The total volume of milk accumulating in the udder of the animals before milking in the Impulse was higher by 0.15 liters than in the UDM–8. But owing to incomplete milking-out of the udder the single milking yield was, on the contrary, lower by 1.42 liters. In comparison with the UDDE–16 the total milk in the udder in the Impulse was higher by 1.55 liters, but the volume of the single milking yield was practically the same owing to different degrees of evacuation of milk from the udder in the two installations. Not only the completeness of withdrawal of milk from the udder but a number of other factors also affect the magnitude of the single milking yields, such as, for example, the productive potential of the cows, the level of fodder intake, and so on. Thus milking machines should not be evaluated on the basis of the values of single milking yields or the 24-hour daily, monthly or complete lactation yields calculated on that basis since this may lead to erroneous conclusions regarding the quality of the milking machines.

The lowest fat content in the single milking yields of cows was recorded in the Impulse type of installation and the highest in the UDM–8. The greater the completeness of stripping the higher the fat content of the milk. However, this fact, though true in general, needs special reservations at present. It is well known that [188] full stripping increases the fat content of the milk collection in cases where enough milk has accumulated in the udder before milking. But if such conditions are lacking in the udder complete stripping does not raise the fat content of the milk collected. It is necessary to examine the indices of fat letdown in cows in milking installations in these terms. These data are given in Tables 17 and 37.

The highest butterfat content in total milk accumulated in the udder at the beginning of milking was in the Impulse installation. Before milking

TABLE 37. FAT LETDOWN IN COWS DURING PHYSIOLOGICAL APPRAISAL OF MILKING INSTALLATIONS

Particulars	Milking installation		
	UDM–8	Impulse	UDDE–16
Amount of fat (g)			
in machine collected milk	236.1	133.8	160.9
in hand stripped milk	35.6	30.6	23.1
in residual milk	130.8	269.6	110.9
in total milk in udder	402.5	434.0	294.9

31.5 g or 7% more butterfat had accumulated in the udders of these animals than in the animals milked in the UDM–8. In spite of this, due to incomplete milking the butterfat content of the milk collected in the Impulse was 1.01% lower and the quantity of fat was 44% less. From these data we can see that even when the potential of butter production was the same incomplete milking reduced the fat percentage of milk and led to a reduction in the amount of fat produced. The indices of fat letdown recorded for cows milked in the UDDE–16 are important in another way. The butterfat content of total milk in the udders of these cows at the start of milking over the entire test period was 0.78% less than in those milked in the Impulse and 0.43% less than those milked in the UDM–8, but the fat content of milk collected (by machine) was 0.7% higher than in the Impulse and only 0.31% less than in the UDM–8. The total quantity of butterfat accumulated in the udder at the start of milking in cows milked in the UDDE–16 was less than in the UDM–8 by 27%, and less than in the Impulse by 32%. But the quantity of butterfat in the milk collected in the UDDE–16 was higher than in the Impulse by 17%. Compared with the UDM–8 it was lower by 32%. These data indicate that in cases where the potential of cows in respect of butterfat production varies from group to group the influence of the milking machine on the completeness of fat letdown may be estimated according to the amount of butterfat in the milk collection expressed as a percentage of total fat content accumulated in the udder at the start of milking.

Which type of milking installation should be preferred on the basis of the data presented in this chapter? Which installation meets the many requirements of the actual workers on the dairy farm, and to what extent? From the results of our tests of milking installations it can be seen that the best conditions for the onset of letdown of milk in cows are created by the UDM–8 type of installation. But this installation is inferior to the 'herringbone' type in point of productivity of labor. Besides this, in the

UDM-8 we do not find some of the necessary conveniences for the operating personnel, above all for the milkmaid, as a result of which, the conditions of work are rather tiring. The milkmaid is placed on the same level as the animal, so that she has to bend while milking. This quickly leads to fatigue on the part of the milkmaid and lowers the productivity of her labor. Apart from this the UDM–8 parlor has a number of design deficiencies such as: an unreliable milk line, inadequate performance of the RVN–100 vacuum pump, which does not provide a stable regime in the vacuum line, and so on. Based on these data it may be said that the UDM–8 meets only some of the numerous design requisites of the dairy industry. So in improving the existing milking parlors and in designing new ones some of the features of the UDM–8 may be utilized: the closed individual type of milking stall and the successful planning of the milking stand, which ensures a certain amount of comfort for the cow. The problem before the designers is to reconcile all these requirements in a unified constructional solution.

The UDDE–16 type of installation is surpassed by the UDM–8 in terms of values of completeness of stripping milk and butterfat. But this installation achieves high productivity of labor. The design offers the maximum convenience to the milkmaid, who does not have to bend while attending to the cows. For proper organization of machine milking in the 'herringbone' arrangement it is very necessary to control the noise coming from the machinery of the milking installation. One good innovation would be to mount a central pulsator with complete sound proofing on the 'herringbones.' It is necessary to maintain maximum silence in the milking installation. Metallic partitions between milking sections on the 'herringbones' should have a soft lining so that when the milking section is closed or opened the partitions do not rattle or clang, particularly where they are constructed from welded tubing. Data on milk letdown in the 'herringbone' show that the construction of the 'herringbones' requires further improvement. It is necessary to pay more attention to the interrelation between the design of milking installations and of milking and auxiliary buildings.

In the testing of the Impulse parlor the data obtained on its techno-operational indices surpassed those of the UDM–8 and the UDDE–16 parlors. High indices of milk quality have also been obtained with this installation. The provision of a sealed milk line and tank in the Impulse ensures milk that is purer and contains fewer microorganisms than in the UDM–8 and UDDE–16. However the design of the Impulse parlor is on the whole unsatisfactory; letdown of milk is severely inhibited because the proper conveniences have not been provided for the animals. But at the same time much convenience has been provided for the milkmaid in the Impulse parlor. In spite of this the Impulse parlor should not be used by dairy farms. The KDU–1 lowers the values of milk letdown as compared to hand milking. Therefore the employment of this installation on dairy farms may lead to a decline in the productivity of cows.

The data presented so far show that none of the milking installations submitted to testing here meets the whole list of requirements in the design of such installations in the practice of dairy cattle farming. And so still more creative work is required on the part of physiologists, engineers, zootechnicians, economists and other specialists before a biologically based technological solution is found for the problems of milking cows in milking installations.

For machine milking, the milking machine M–59 should prove to be of considerable interest. Compared with the DA–3M and the DPR–2 it ensures a more complete stripping of milk and fat from the udder at high speed without any need of manual stimulation of milk letdown. The DPR–2 is excelled by the M–59 in the completeness of withdrawal of milk and fat from the udder while the DA–3M yields to the M–59 in speed of milking as well as completeness of milking.

The DA-3M is the improved version of the DA–3. The big improvement on the DA–3 is in the reduced duration of the cycles of compression and rest and prolongation of suction. The ratio of time between strokes in the DA–3 is as follows: suction 45, compression 15, resting 40. In the DA–3M it is 60–10–30. Until recently our industry manufactured only the DA–3. It has been withdrawn from production now but on many of our dairy farms cows continue to be milked by this machine. The completeness of stripping of milk by the DA–3 was tested with 15 cows in the sovkhoz "Krasnaya Poima." These cows were already being milked with the DA–3. They were not specially selected for the test but all the cows taken for the test were being served by one milkmaid. The results of this study are given in Table 38.

As we can see from these data the DA–3 could not withdraw as much milk as was drawn by hand owing to its very slow speed of milking. With such a low milking performance it was rather unreasonable to expect an improvement after its introduction on the farm. The rate of milking of the DA–3 is not much more than one-third that of the M–59. The DA–3 cannot withdraw milk coming in by the letdown reflex fast enough and this in turn inhibits the outflow of milk from the alveoli. This should explain the large volume of the hand strippings in Anikin's experiments [8, 9]. He tried to organize fast machine milking of cows with the DA–3 with the vacuum increased to 50 cmHg. The sole advantage of the DA–3 appears to be its safety in respect of outbreaks of mastitis.

Modern milking machines should be able to withdraw milk rapidly from the udder, the time taken should be short, and the speed of milking high. The milking machine should completely withdraw the milk flowing

TABLE 38. MILK LETDOWN IN COWS MILKED BY MILKING MACHINE DA–3

Particulars	Data
Duration of milking (min sec)	5.08
Rate of milking (ml/min)	570
Milk obtained by machine (ml)	2,930
–Do– as %	62.0
Hand strippings (ml)	887
–Do– as %	18.6
Residual milk (ml)	915
–Do– as %	19.4
Total milk in udder (ml)	4,732
–Do– as %	100

out of the alveoli owing to the letdown reflex to the teats and cisterns of the udder. Where the device is able to draw out completely the milk entering the teats as a result of letdown there need be practically no hand stripping at all. But it is to be regretted that this requirement is as yet not satisfied by any milking machine. Analyzing the indices of two-phase milking machine, Espe [61] writes that after the cow begins to 'give up' her milk the mechanical means of milking employed are unable to withdraw milk from the udder at the speed at which it enters the udder cisterns.

The milking machine should exert a stimulating effect on milk letdown, i.e. it should bring about a very active letdown reflex or else reinforce the reflex if it has already set in response to the preliminary treatment of the udder. Thereby more milk should issue from the alveoli and there should be less residual milk in the udder at the end. Therefore among the different milking machines the best one would be that which milks out the cow most rapidly leaving the minimum volume of hand strippings and residual milk. Of the three machines tested in our present study the best in these respects is the M–59. This milking machine should be more widely introduced on dairy farms, kolkhozes and sovkhozes.

How does the M–59 manage to stimulate the unconditioned reflex of letdown to a larger degree than the DA–3M and DPR–2? The M–59 milks out the teats in pairs and has an alternating work regime which is as follows: if the fore teats are in the compression phase then the rear teats are in the suction phase. As we know, in hand milking usually two quarters are milked, the front ones first and then the hind ones or maybe in the reverse order. More than this, in hand milking the alternating regime of milking obtains even for the same pair of teats. In milking a pair of teats the compression of the teat and squeezing out of milk involve first one teat, then the other. Hand milking of cows has been practiced over many millenia and this alternating regime of milking the udder has apparently become a habitual stimulus for the reflex activity of the cow. It may be presumed that the M–59 with its alternating regime of work to some extent reproduces this habitual stimulus for the udder receptors which initiate the unconditioned reflex letdown in the cows. In any case, the working principle of the M–59 is worthy of attention and should be taken into account in the elaboration of a better design for milking machines.

It would not be proper to explain incomplete stripping of cows solely by the deficiencies of milking machines and installations, though this may be the main reason for the low efficiency of machine milking on dairy farms. In our tests of milking installations, regardless of type, at some milkings all three types of machine tested withdrew all the milk from the udder. Results of these experiments are given in Table 39.

As we have already seen, in the machine milking of cows, there is an immense potential capacity for inducing milk letdown. This could make possible the achievement of very high levels of withdrawal of milk from the udder. It is to be regretted, that among the many specialists concerned with the design of milking apparatuses there are some inaccurate conceptions on the physiological capacity of letdown of milk in cows. Korolev [99] thinks, for example, that even with the best methods of milking some 3–4 liters of so-called residual milk remains in the udder. But this view is not confirmed by experiments.

Is it at all possible to secure full evacuation of the quarter of the udder by milking machine? Data on the yield of hand-stripped milk and residual milk obtained after machine milking from each quarter of the udder are given in Table 40.

As we can see, under conditions of normal milking it is often possible to withdraw all the milk from the quarter of the udder in milking installations. This attests to the fact that in milking installations as well as in other modes of milking the full evacuation of the quarter of the udder is a normal physiological phenomenon. However, the causes underlying the retention of some milk in the quarter of the udder, as well as the nonuniformity as between the four quarters, are as yet not fully understood. Research on the causes of irregular stripping of milk from the quarters in machine milking is of great importance, especially in regard to delay in milk letdown and the hardening of the udder quarters in relation to the prevention of mastitis in cows.

In evaluating milking machines it is usually assumed that the faster a machine strips the milk from the udder the better its design. This view, which is correct in general, requires some reservation at present since the raising of the milking speed does not always lead to fuller evacuation of the udder by the milking machines. The milking speed of the UDDE–16

TABLE 39. MAXIMUM VALUES OF EVACUATION OF THE UDDER BY MILKING MACHINES WORKING IN DIFFERENT TYPES OF MILKING INSTALLATIONS

Particulars	Dates of experiment (1961)						
	Aug. 2	Aug. 11	Aug. 11	Aug. 14	Aug. 17	Sep. 7	Sep. 7
Type of installation	UDDE–16	UDM–8	Impulse	UDM–8	Impulse	UDM–8	UDM–8
Milking machine	DPR–2	DA–3M	M–59	DA–3M	M–59	DPR–2	DPR–2
Serial number of cow	8,662	6,018	376	6,656	1,426	6,656	8,912
Milk withdrawn by machine (ml)	4,000	5,000	3,500	7,600	7,200	6,000	5,300
–Do– as %	93.6	94.5	96.0	94.7	93.8	94.5	93.4
Hand stripped milk (ml)	75	100	50	150	75	50	200
–Do– as %	1.7	1.8	1.3	1.9	1.0	0.8	3.5
Residual milk (ml)	200	200	100	275	400	300	175
–Do– as %	4.7	3.7	2.7	3.4	5.2	4.7	3.1

TABLE 40. DISTRIBUTION OF HAND-STRIPPED AND RESIDUAL MILK IN THE QUARTERS OF THE UDDER AFTER MACHINE MILKING (IN ML)

Sl. No. of cow	Particulars	Udder quarters			
		Right front	Left front	Right back	Left back
8662	Hand-stripped milk	—	—	35	50
	Residual milk	—	50	50	100
6018	Hand-stripped milk	50	—	50	—
	Residual milk	100	—	100	—
376	Hand-stripped milk	—	—	20	30
	Residual milk	—	—	50	50
6656	Hand-stripped milk	—	50	100	—
	Residual milk	—	75	100	100
1426	Hand stripped milk	75	—	—	—
	Residual milk	200	100	—	100
8912	Hand-stripped milk	—	100	50	50
	Residual milk	—	—	—	175
6656	Hand-stripped milk	—	—	50	—
	Residual milk	—	—	300	—

was higher than the UDM–8's but the completeness of stripping milk and fat from the udder was lower. Installed in the UDM–8 the DA–3M and DPR–2, working at different milking speeds, withdrew similar amounts of milk and fat from the udder. The M–59 and the DPR–2, which have the same milking speed, withdrew different quantities of milk. All this shows that the speed of machine milking cannot be used as an independent index for evaluation of the performance of milking machines. At the same time, assessment of the speed of milking in relation to completeness of withdrawal of milk and fat from the udder would allow a more precise approach to the evaluation of the quality of machine milking: the better apparatus is the one that is not only fast but also milks out the cow more completely.

In judging the data of physiological testing in relation to milking plants the question may arise whether the differences in milk letdown in different kinds of milking plants are due less to the installations than to differences between the groups of cows tested. This question would be relevant under conditions where the cows tested in the UDM–8, say, have a high level of udder evacuation, those in the UDDE–16 have a lower level and those in the Impulse have an inherent tendency to retain milk in the udder. But the letdown of milk in all the experimental groups of cows was characterized by fluctuations from one milking to another. Any assumption that there could have been different inherent levels of evacuation of the udder in different groups of cows just cannot stand up. Among cows there are differences in regard to the completeness of withdrawal of milk from the udder. One animal may hold back in the udder more residual milk and another less. Based on these data, it is possible to surmise that in the UDM–8, say, a few cows will hold back a large quantity of residual milk, in the UDDE–16 more cows will do so and in the Impulse still more. However, in our physiological testing of milking installations we did not get this kind of data. In comparing the milk letdown indices of cows in the three plants, the UDM–8, UDDE–16 and Impulse it was found that for each cow the quantity of withheld milk ejection over the test period was lowest in the UDM–8 and highest in the Impulse. As a result the quantity of residual milk in each cow in the Impulse turned out to be higher than in the UDM–8. It may be said in general that in cows there exists no such thing as a biologically inherent capacity to retain a larger or smaller quantity of residual milk. Retention of residual milk occurs due to inhibition of the milk letdown reflexes. So individual peculiarities of the cows may not exert any significant influence on the indices of milk letdown in the evaluation of milking machines. But, unfortunately, until very recently undue attention was paid to the problem of creating individualized conditions for the cows while scant attention was paid to the problem of working out standardized conditions of exploitation of the animals, particularly to the biological basis of standardization of methods of milking, handling and maintenance.

A clear concept of the role of the relationship of conditioned and unconditioned factors in the process of milk letdown in cows is of great importance for a scientifically based organization of machine milking of cows. The milking installation with all its complex of surrounding conditions molds the conditioned reflex reaction of the cow to being milked. In the UDM–8 the environment for the onset of milk letdown was the most favorable. As a result the cows were better prepared to be milked. But to obtain high values for withdrawal of milk from the udder in a milking plant under conditions where conditioned reflexes stimulate milk ejection it is very necessary to have more indirect influences promoting the unconditioned reflex of milk letdown. In the UDM–8 the use of unconditioned stimulation like udder massage reinforced reflex milk letdown in cows when they were milked by the DA–3M. When these same cows were milked in the UDM–8 by the M–59 the letdown of milk was intensified. On the basis of these data we are entitled to assume that the conditioned reflexes of milk letdown are fundamental for a more effective realization of direct milking influence that promote the unconditioned reflex of milk letdown.

At the same time, in cases where direct milking influences evoke an unconditioned milk letdown reflex of low intensity, even the presence of a conditioned reflex predisposition on the part of the cow toward machine milking may not ensure high indices of completeness of stripping. It is probable that the mere presence of the conditioned component of milk letdown is insufficient for maximum evacuation of the milk from the alveolar spaces of the udder.

On the other hand the absence of any predisposition on the part of the cow to milking and, consequently, the absence of conditioned reflexes of milk letdown, lead to this situation: the direct milking influences which in favorable conditions usually stimulate the unconditioned milk letdown reflex (as in the UDM–8), exert no influence on milk letdown (as in the Impulse). Also, the conditioned and unconditioned components of milk letdown should not be considered as independent processes which may lead separately to the same completeness of udder evacuation. Thus in the practice of machine milking improvement of the complex of external conditions in which milk letdown takes place, which means the improvement of milking installations, is as important as the modernization of the milking machines, which are essentially responsible for the unconditioned reflex of milk letdown.

Chapter V

Physiological Requirements in the Design of Milking Installations

Constituents of the dairy premises

Dairy buildings consist of a milking parlor and ancillary areas. The latter are of two kinds—incoming and outgoing. The vacuum pumps, milk tanks and other equipment are installed in outbuildings. The milking parlor should provide normal conditions for milking of cows. The entrance zone of the dairy building serves to prepare and condition the cow to be milked. It should have a durable floor with a firm surface and be provided with good sewerage and a system for washing away cowdung with water. Sometimes instead of an entrance zone there may be a long, narrow passage which the animals are sent down to be milked. Before the entrance to the milking parlor, some dairy farms have special stalls for washing the cow's udder. But this would work against the natural course of the milk letdown reflex. The section through which the cow enters the milking stall should be situated on the same level as the floor of the milking stall. The enclosure of the entrance section should be rounded in form. The length of the entrance section should not exceed 2–3 m. This section should be designed to help the cows relax before milking. Though this aspect has not been studied closely it may tentatively be recommended that a minimum of 3 m^2 of floor space per cow be provided in the entrance section of the premises. The exit section serves for assembling groups of cows after they are milked.

Construction of the milking stall

The interior of the milking stall should be designed to give the cow maximum comfort. The stall should be large enough to allow the cow to stand without any constraint. A rectangular form of milking stall has been adopted in recent years. But this form does not provide the necessary comfort to the cow. The design of the interior of the milking stall should use curves. This would provide better conditions for the onset of milk letdown in cows. A milking stall without sharp angles not only precludes traumatic injuries and bruises to the projecting parts of the cow's body but also ensures minimum excitation of the cow when she enters it, during her stay there and exit.

The quality of the material of which the individual components of the milking stall are constructed also has importance. The material used for making the stall should have resilience. It is well known that in planning dairy installations designers more than anything else aim to provide durability of construction in building the milking stalls. But this zeal for durability may lead to unnecessary rigidity of parts likely to come into contact with the cow's anatomy. It is natural that the animals respond to this kind of milking stall with inhibition of letdown of milk. Therefore construction of the components of the milking stall from materials having not only durability but, above all, resilience, permits favorable conditions for milk letdown in cows. Unfortunately, the materials nowadays used for milking stalls (wood, asphalt, concrete) do not meet these requirements. They possess neither durability nor resilience. But at the same time materials like asphalt and concrete acquire even more rigidity with use and this causes rapid wearing down of the underside of the cow's hoof. The growth of the horny base of the hoof in these conditions cannot compensate for the wearing down of the horn. As a result, the animals often acquire a limp and this leads to a lowering of productivity. Sometimes rapid wearing down of the horny hoof when the cows are kept on concrete floors does not cause visible limping but the animal moves with a gingerly gait. All her attention under these conditions is on ensuring painless movement of the body and securing for herself a comfortable position in the milking stall. The reaction to milking after this experience will be one of inhibition. Therefore a detailed experimental analysis of such "trifles"—as they appear to be at first glance—will have much practical importance in the proper planning of the entrance area and milking stall.

Shape of the milking stall

In analyzing the data on the completeness of milk letdown of cows in installations of the Impulse type it was observed that the shape of the milking stalls here does not offer the required comfort for the cows as she takes up her position. What are the requirements of the shape of a milking stall? We may ask. The ideal form would be one that allowed for proper positioning of the cow's body in the stall, such that the walls did not come into contact with the projecting parts of her anatomy. Therefore the shape of the milking stall should strictly conform to the spatial requirements of the cow's body and offer the maximum comfort to the animal. These are factors that promote the generation of a stable positive reaction on the part of the animal to the installation.

In order to guard the cow against distractions when she is in the milking stall the design of the stall should be of the enclosed or semienclosed type. The milking parlor is a very suitable object for automation of the milking process. It is comparatively easy to develop in the animals the conditioned reflexes of sound stimuli. But the use of this stimulus is complicated because the sound signaling of productive processes leads to superfluous sound effects. In milking stalls of the enclosed type the generation of the conditioned reflex to light stimuli acquires considerable importance. The linking of light stimulus with the accomplishment of some operation in the preparation of the cow for milking or in actual milking will be conducive to the development and gradual consolidation of conditioned reflexes to the milking process, and this in the end leads to the stimulation of the milk letdown reflex. This would allow the standardization of the milking operation to a certain extent and this in turn would facilitate the use of various automatic systems in milking installations. It is more difficult to generate conditioned reflexes to light stimuli than to sound stimuli [27]. Hence enclosed milking stalls facilitate the generation of conditioned reflexes to light stimuli. Considering the fact that cows can distinguish and evince conditioned reflexes to red, green and blue lights in enclosed stalls, it is possible to connect a number of automatically executed operations during the milking process with these colors.

Basic requirements of the working parts of milking installations

In the technology of continuous milking, the reliability of the working parts of the milking installation is of great importance. The breakdown of machinery in the milking installations immediately disrupts the daily routine of large numbers of cows and this in turn leads to a lowering of milk production and other undesirable consequences. Some 12–20 cows are milked at a time in an installation, depending on its design. If the vacuum pumps, for example, fail at milking time the teat cups fall off and the reflex of milk letdown is disrupted. Since it cannot be evoked again the animals remain unstripped. With a single milking yield of 5 liters this will lead to a loss of milk output per herd of 60–100 liters. The negative effect of frequent breakdowns of machinery in the milking installation is not confined to this. The disruption of the habitual rhythm in the animal leads to the derangement of reflexes generated in the milking installation and this may lead to delay in milk letdown at subsequent milkings when all the machinery of the installation is working normally again. So in order to guarantee the normal working of the milking installation it is necessary that the machinery be duplicated so that an automatic system switches on the standby machinery the moment there is a breakdown. The provision of standbys for the most important mechanical systems and the provision for switching them on automatically ensures reliability and normal regime of work in milking installations. In installations getting their power from the electrical grid it is necessary to have a reserve electric power plant with automatic switches. This should pick up within 15–30 seconds of the failure of the main line. This averts the dropping off of the teat cups.

Maintenance of a steady vacuum regime in the teat cups, which depends on the work regime of the pulsators, collectors, milk line and vacuum pump, is a very important prerequisite which ensures normal conditions for milking cows in dairy installations. Instability of the work regime of these systems and mechanisms is at present the most vulnerable part of the technology of machine milking. All mechanisms in the installation should work noiselessly. Soundproofing and muffling of sounds coming from feed conveyors, motors, pulsators, milking machines and other mechanisms working in the installation is of great practical importance.

The milking machine exerts a direct influence on milk letdown in cows. The stimulation of the unconditioned letdown reflex and the speed of withdrawal of milk from the udder depend on the quality of the milking apparatus. An examination of a few prospective improvements in milking machines is of great interest for the practice of machine milking of cows.

Speed of machine milking of cows

Rapid stripping of milk from the udder depends not only on the milking machine but also on the cow. For the speedy milking out of the udder there should be reciprocal action between the milking machines and the cow. But is the cow in a position to give up all her milk in the course of two-three minutes? Trials [183] have given an affirmative answer to this question. On the other hand, is the milking machine in a position to milk out from the udder in two-three minutes all the milk the cow can give up as a result of the milk letdown reflex? In a large number of tests on the rate of milking by milking machines it turned out that in some individual milkings the rate of milking reached 4,788–5,930 ml/min [23, 51, 66, 73]. In two minutes of milking at this speed—if the cow yields milk at this rate—the milking apparatus will be able to milk out 8–12 liters of milk. These data show that with machine milking there are great possibilities of raising the speed of stripping of milk from the udder. The machine stimulates unconditioned reflex milk letdown in the cow. Naturally the activity of the unconditioned reflex of milk letdown depends on the strength and nature of the action of the milking machine on the nerve endings of the teats.

Effect of the frequency of teat compression on milk letdown

For the onset of the milk letdown reflex it is very necessary for the nerve endings of the walls of the teats to be stimulated so that they send impulses to the cerebral hemispheres, from where signals go out to the smooth musculature of the udder, causing it to relax, and to the pituitary, which releases oxytocin. This reflex can appear only under certain magnitudes of stimuli, i.e. magnitudes of frequency and strength of teat compression, at milking time. But the onset of the milk letdown reflex in cows in response to certain stimuli cannot characterize the end result of the letdown—the completeness of withdrawal of milk from the udder. It is very important to ascertain at what frequency of teat compression all the milk can be withdrawn from the alveoli of the udder. Many frequencies of teat compression (also known as pulsation rate) ranging from 45–120 per minute are used in milking machines. But these rates are not empirically based. They have been introduced into the design of each apparatus on the basis of data obtained from the study of hand milking or of the sucking of milk by calves. In some experiments [75, 181] the effect of frequency of teat compressions on milk letdown is characterized according to the volume of single milking yields. But a number of factors exert their influence on single milking yield and, therefore, this index by itself is not helpful for such purposes.

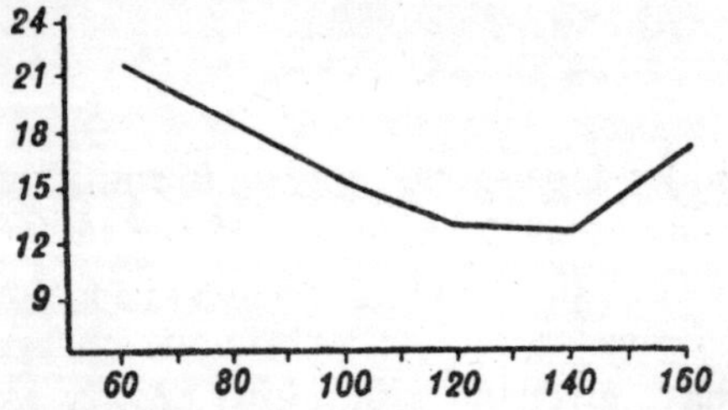

Fig. 10. The effect of the frequency of squeezing teat on the completeness of milking of a quarter.

On the vertical axis—*residual milk in % of the entire volume of milk in the quarter;* On the horizontal axis—*number of squeezes per minute.*

In the Grachev's experiments [70] on goats the cisternal milk was drawn out of the quarters by catheter and the teats were then stimulated manually at a known frequency and the duration of the latent period of the milk letdown reflex as well as the volume of the alveolar portion of the milk were determined. According to Grachev's data the optimal frequency of teat compression for evoking the milk letdown reflex in goats is 84–132 compressions per minute. Unfortunately the residual milk was not taken into consideration in these experiments. This factor complicates the evaluation of these data. Besides, this wide range of teat compression frequencies able to evoke the milk letdown reflex in goats does not allow the determination of the precise frequency to be employed in milking machines to strip the milk more completely from the udder. The need to check Grachev's data is illustrated by Marchenko's experiments [119], where 25% of total milk was retained in the udder of cows milked with a teat compression frequency of 80 per minute while only 17% was retained with a frequency of 140. It is true, of course, that the method employed by Marchenko of obtaining milk from the entire udder is not precise enough. The quarters milked were not stripped simultaneously by hand milking and this factor may have had a great influence on the volume of residual milk.

In view of the fact that the teat compression frequency is one of the important factors governing operation of the unconditioned milk letdown reflex, we studied the influence of teat compression frequency on the completeness of milk letdown [185]. The experiments were carried out on three cows with a daily milk yield of 17–23 liters. The animals were milked three times a day by hand. To eliminate conditioned reflex influences associated with the act of milking, the studies were carried out one hour before the daily milkings beginning 7 hours after the previous milking. The frequencies of teat compression employed in hand milking by the milkmaid were 60, 80, 100, 120, 140 and 160 compressions per minute. In conformity with these frequencies the experiments were organized in six series. In each series involving three cows 15 experiments were run, giving 90 trials over the period of this project. One experiment was run with each cow everyday. A plexiglass catheter was introduced into the right rear teat and the cisternal milk withdrawn. Then after 10 minutes the milkmaid took over and in the course of two minutes, by exerting the required pressure and frequency, stimulated the right front quarter of the udder and withdrew the cisternal milk. After the flow of the alveolar portion of milk through the catheter in response to the stimulus had stopped, an injection of 18 IU of oxytocin was given in the jugular vein and the residual milk stripped by hand from each quarter into a four-chambered milk pail. The completeness of milk letdown was ascertained in the following way: The cisternal, alveolar and residual milks were assumed as 100. The residual milk was expressed as a percentage of the total volume of milk of the quarters. The results of these trials are given in Fig. 10. The milk letdown reflex appears in cows under all the teat compression frequencies tried. The greatest milk letdown activity was observed with a 120–140 per minute teat compression frequency, where the smallest quantity of residual milk was retained in the alveoli. These data may be considered to be approximate, since with hand milking it is scarcely possible to measure precisely the teat compression frequency. But in spite of this it is quite apparent that raising the frequency of stimulation of the udder teat receptors hastens the milk letdown reflex in the cow and this leads to a more complete evacuation of milk from alveolar spaces. Of course, further study of the effect of teat compression frequency on the completeness of emptying the milk from the alveoli cannot be conducted without the instruments needed to measure the required teat compression frequency accurately and provide the cow with the stimulation to which it is habituated in hand milking.

Several researchers have observed that when the number of pulsations in the milking machine increased the rate of milking also increases [6, 22]. Milking machines were designed with a pulsation rate of 120–140 per minute [58, 145]. But it turned out that the completeness of udder evacuation by the DPR–2 milking machine with a pulsation rate of 120 per minute was the same as by the DA–3M with a pulsation rate of 60 per minute. The M–59, with the frequency of 60 pulsations per minute, strips the milk from the udder more completely than the DPR–2. What could be the reason for this discrepancy? It may be supposed that the change in the number of pulsations in the milking apparatus leads to a change in other parameters—teat compression force, magnitude of teat cup vacuum, duration of phase of compression of teat cup liner, and so on. It is true that in one experiment [85] where the pulsation rate was increased from 30 a minute to 90 the completeness of udder evacuation also increased, but these data need checking because residual milk was withdrawn in this experiment with the help of 10 IU oxytocin. At a teat compression frequency of 120–140 a minute, 86–87% of the milk is withdrawn from the udder, but the M–59 and DA–3M can also withdraw 85–86% of the milk with 60 pulsations per minute. The findings of these experiments indicate that other parameters of the milking device also exert an influence on the activity of the unconditioned reflex of milk letdown. The study of the degree of influence exerted by each parameter of the milking machine on the activity of the unconditioned reflex of milk letdown will enable us to incorporate in the milking machine the complex of unconditioned stimuli which will ensure maximum evacuation of milk from the udder. The data obtained from such applied research will constitute in the final analysis the contemporary physiological basis for the design of improved types of milking machine.

Effect of teat compression force on milk letdown

Among the factors that may have an effect on milk letdown teat compression force is of great importance. In the walls of the teats there are a large number of complex nerve endings enclosed in thick capsules [30, 120]. These receptors respond only to high indices of pressure on the tissues. It is no coincidence that they innervate the mesenteric muscular tissues, the horny pad (in the jaw) and other tissues that are subjected to mechanical actions and pressure. According to Peinovich [141] the force of teat compression exerted by the milkmaid's hands attains 53.4 cmHg, during which the individual fingers produce the following pressures: index 20.5, middle 14.6, fourth finger 12.1 and little finger 6.2 cmHg. The compression force on an artificial teat is easily ascertained manometrically. In large-scale measurements of the teat compression force that we conducted it turned out that a very highly skilled milkmaid often milks the cow with a pressure on the teats equivalent to 70–76 cmHg (total for all fingers) with the force applied by the index finger sometimes reaching 25 cmHg.

Therefore study of the pressure on the teats is of great practical significance. To study the effect of teat compression force on the activity of the milk letdown reflex in cows we employed the same method as for studying the effect of teat compression frequency on milk letdown. The following levels of pressures on teats were studied with three cows: 7.6, 22.8, 38.0, 53.2, 60.8 and 76.0 cmHg. Before each experiment the milkmaid thoroughly checked the required teat compression force on a dynamometer and then applied this stimulus on the teats. The results of this research are given in Fig. 11. As we can see, the quarters of the udder are most completely evacuated of milk at teat compressions of 53–60 cmHg. Further increasing the pressure on the teats does not increase the intensity of the milk letdown reflex. Teat compression force has a much greater effect on the intensity of the milk letdown reflex than teat compression frequency. At optimal magnitudes of teat compression force (53–60 cmHg) the completeness of udder evacuation reaches 88–89%. These indices are higher

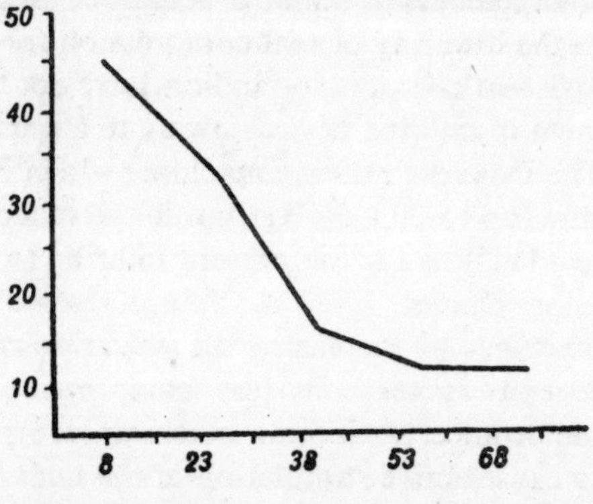

Fig. 11. Effect of teat compression force in milking on completeness of evacuation of milk from the quarters of the udder.

Vertical axis—*residual milk as percentage of total milk in the quarters;* Horizontal axis—*pressure in cmHg.*

than those obtained with the M–59 and DA–3M. It may be mentioned here that hand milking of average efficiency, as shown by measurement* involes teat compressions of 30–45 cmHg, and of poor quality, 20–28 cmHg. This pressure is hardly sufficient to overcome the resistance of the teat sphincters, which in hand milking reaches 20–31 cmHg [141] and in machine milking 10–30 cmHg [100,112]. Milk letdown with this pressure is weak and as a result of this 30–35% of the milk is held back in the alveolar spaces.

*Twenty milkmaids participated in the manometric measurement.

For a proper understanding of the nature of the unconditioned milk letdown reflex in cows it is very important to explain the role of some specific factors that conduce to stimulation of the teat receptors: for example, is the stimulation of the receptors produced by pulling of the teat or is the stimulation produced by compression? Tsakhaev [173] thinks that the stretching of the deep tissues of the walls of the teat plays the chief part in arousing the milk letdown reflex. It is difficult to correlate the published factual material with this view. When the udder is filling up in the interval between milkings the teat walls are stretched but this does not bring about the milk letdown reflex. In first-calf cows [192] the pressure in the udder between milkings may reach 70–75 cmHg. The teat walls are stretched but milk letdown does not occur as a result of the stimulus of intra-udder pressure. But if we start compressing the teat with a pressure of 6–8 cmHg the milk letdown reflex is activated in response. These data show that it is not the pulling but the compression of the teat walls of the udder that plays the chief part in stimulating the milk letdown reflex in cows.

To obtain the milk letdown reflex in cows it is necessary to employ high indices of teat compression but it has been observed that in machine milking of cows with machines of all makes the teat compression falls short of the indices that cause milk letdown of high intensity. It follows that a mere increase in the pulsation rate in the milking machine has no effect on completeness of udder evacuation. In fact raising the pulsation rate leads to the lowering of teat compression force. It may be mentioned here that high teat compression indices have not been included in the design requirement of milking devices owing to a number of serious technical difficulties. The Doyarka milking machine, which is designed to maintain a high pressure (up to 30.4 cmHg) inside the teat cup liner to squeeze milk out of the teat [142], is for the present inferior to other machines and needs further improvement. Even so, future research on the improvement of milking machines by increasing the pulsation rate to 120 per minute with a simultaneous increase in teat compression force will be of great practical importance. A milking machine incorporating these features will be able to cause active stimulation of the unconditioned milk letdown reflex.

Pulsation ratio and teat cup vacuum level in milking machines

The ratio between the phases of each pulsation (suction and compression in two-phase apparatuses and suction, compression and rest in 3-phase apparatuses) has a great influence on the milk letdown reflex. Raising the suction phase in 3-phase milking devices from 45 to 60% of the duration of the pulsation [115] increased the rate of milking [74]. Research with ratios between the suction and compression phases varying from 1 : 1

to 5 : 1, and the suction phase filling from 50 to 84% of the time, showed that this increased the rate of milking [40, 65, 144]. But the rate of milking cannot characterize the intensity of the milk letdown reflex and so these data do not serve to substantiate the influence of the pulsation ratio on the intensity of milk letdown. In the M–59 milking machine the suction phase lasts longer than in the DA–3M, with the result that in the M–59 the speed of milking and hence the completeness of stripping of milk from the udder are greater. But in the DPR–2 the duration of the suction phase is greater than in the DA–3M but the completeness of udder evacuation is the same. All this shows that the question of the effect of the pulsation ratio on milk letdown requires further research.

The teat cup vacuum level may also influence the intensity of the unconditioned milk letdown reflex. When the teat cup vacuum is raised 25 to 51 cmHg there follows an increase in speed of suction of milk from the teats [21, 38, 113]. However, the effect of this factor on the completeness of udder evacuation has not been studied. The magnitude of the teat cup vacuum level adopted these days in milking machines is determined first of all by consideration of the safety of the cows milked by machine [126, 127]. High milking speeds can be obtained with high teat cup vacuum levels. However, the adoption of high teat cup vacuum levels in milking machines requires the incorporation of protective devices to ensure the safety of the udder during milking.

Physiological basis of the theory of the design and construction of milking machines

For the formulation of a mathematically based theory of milking machine design it is necessary to establish what conditions ensure complete evacuation of the udder in milking by machine. In other words, it is necessary to know which parameters of machine milking, such as teat compression force and frequency, pulsation ratio, magnitude of vacuum, direction of teat compression, deformation of teat cup liner—are necessary for complete udder evacuation. Besides this, it is necessary to make a further thorough study on the degree of fullness of the udder at which the various parameters of the milking machine will be effective as intended. To obtain such basic data it is necessary to carry out physiological research in the form of studies on the milk letdown phenomenon in cows and the factors (in this case the parameters of the milking machine) that exert an influence on this phenomenon. What should be the physiological basis of the design of milking machine? The working principles of contemporary milking machines, 2–phase as well as 3–phase, are quite different from hand milking [185] and the suckling of calves [122]. In conducting research on applied physiology with a view to improving milking machines the process of hand milking or suckling of calves need not be imitated but we have to find out by precise experiments the factors necessary for complete udder evacuation. After milking by machine some 24% of residual milk on an average is left in the udder. The solution to the problems of improvement of milking machines centers on the method of withdrawal of this residual milk. If a milking machine were to withdraw all the milk from the udder together with the 24% that constitutes the residual milk and is now absorbed into the daily milk yield, other things being equal, the milk productivity of the cows would naturally increase by 24%. Over and above this, complete stripping of the milk from the udder opens up possibilities for great expansion of milk yields in machine milking by increasing milk formation in the udder following each complete stripping of milk. There are many possible approaches to a solution of the problems mentioned. Korolev [100] thinks that the simplest way to stimulate the letdown of alveolar milk in machine milking is the use of mechanical massage. According to him it would be very expedient to stimulate milk letdown at the time when the cisternal milk is being milked out. The udder could be milked and massaged simultaneously. Massage could be continued over the entire duration of milking so that the letdown of milk did not slacken. Based on this idea, a number of different types of mechanical massagers for cows and different kinds of milking machines with milk letdown stimulators were built in Korolev's laboratory. It must be recognized, however, that the use of massa ers would complicate the design of the milking machine. Mechanical massagers have not yet been subjected to physiological testing of their action. Meanwhile the expediency of using them will be justified if they promote a fuller evacuation of milk from the udder as compared to milking machines without massagers.

Hand massage of the udder after putting the teat cups on the teats stimulates milk letdown and facilitates a fuller evacuation of milk from the alveolar spaces. But the same indices of completeness of stripping of milk from the udder in machine milking can also be obtained without massage of the udder. These data cast some doubt on the prospect of employing udder massage to stimulate milk letdown in machine milking. The foregoing does not rule out altogether the utility of mechanical udder massage to stimulate other functions, for example, in the development of the tissues of the mammary gland. But for such purposes massagers of a different type are required, based on the principle of stroking the skin of the udder with mechanical bristles at relatively low pressure [7]. The massaging device in this case will be an apparatus working independently. Massaging is recommended for udder development in calves, in cows in their dry period and in the intervals between milking to stimulate the development of the capacity of the udder and tone up its functions [184]. Therefore the designing of miniature mechanical massagers is of promise in the mechanization of work for the development of the udder cavities and for milk yield increase.

In the tissues of the teats of the cow's udder there are receptors whose stimulation governs the unconditioned reflex action of milk letdown. Therefore the reflex zone of the teats is responsible for the unconditioned reflex function of milk letdown and completeness of evacuation of milk from the alveolar spaces of the udder. So we have to proceed in other ways for the improvement of milking machines. It is very necessary to research technical procedures which would reinforce the stimulation of the teat receptors. From this it follows that the design of teat cups, collectors and pulsators should be the main object. Their improvement would ensure intensive stimulation of the teat receptors, which the animals would get used to. This would result in raising the level of milk letdown reflex activity in machine milking. Thanks to this it would be possible to withdraw all the milk from the cow's udder by machine at every milking. The need for additional stimulation of milk letdown would disappear and the process and technology of machine stripping of milk would be simplified.

The data presented in this chapter show that the physiological requirements confronting the designers of milking installations comprise three basic elements: 1) requirements of the milking hall and ancillary premises, 2) requirements of design of milking parlor, and 3) requirements of the design of the milking machine. The improvement of the technology of machine milking depends in the final analysis on the interrelated and comprehensive development of all these three elements of dairy technology. Inadequate design of this or that element, for example, the milking premises, may impair the results of work on the improvement of the other two elements—the milking installation and milking machine. These aspects of the development of the new technology of machine milking should be kept in mind to avoid lowering the milk productivity of the cows. New milking techniques should not be used on dairy farms without physiological checking. Premature introduction of a range of dairy equipment whose use inhibits milk letdown in the animals is economically unprofitable. This calls for the reorganization of all aspects of work related to the development, improvement and introduction of milking machines.

Chapter VI

Physiological Basis of the Technology of Milking Cows in Milking Installations

Feeding regime of cows

The correct feeding of cows facilitates their preparation for milking. Their allowance should include the supply of nutritional elements necessary for the formation of milk and also a certain amount of succulent roughage. The feeding regime of bulky feeds should ensure uniform distribution over the course of the day. There is a functional relationship between the mammary glands and digestive organs in cows. The stimulation of the mammary gland receptors increases the contractive activity of the rumen [159] and causes reflex chewing of the cud [68, 138]. On the other hand, experiments on cows [162] and goats [72] show that impulses originating in the digestive system may have an influence on milk secretion. In Sinesh-chekov's laboratory [163] it was shown that milk yield was higher with four-times-a-day feeding than with twice-a-day feeding. But the milk yield also varied with the completeness of stripping of milk. Hence it was necessary to establish how the feeding regime of cows affects completeness of stripping of milk from the udder.

Experiments were carried out on five cows of the Red Steppe breed with a daily milk yield of 18–26 kg. The allowance consisted of 3 kg lucerne hay, 25 kg corn silage and 3 kg concentrates. The feeding regime was as follows: Silage was fed three times, once after each milking. The hay was fed twice after the morning and noon milkings. The concentrate was given after the noon milking. The experiment was run over periods of nine days each over a stretch of 45 days. During the first period of the experiment (control), the feeding regime was not changed. In the subsequent periods the adopted ration was fed to the cows over 11 hours. The cows received nothing for 13 hours in the next period. In the third period they were fed three hours before the morning milking with 1.7 kg concentrate as a supplement to the ration, in the fourth period with 9 kg of silage and in the fifth with 4 kg of hay. The nutritive value of the supplements is shown in Table 41. The cows were hand-milked three times a day: at 6 in the morning, noon and 8 in the evening. Milk letdown was reduced at the morning milking once every three days, i.e. three times in the course of every nine-day period. The residual milk was stripped after injection of 18 IU of oxytocin into the jugular vein. During each period 15 trials were carried out on five cows. In Table 42 we give the mean data of the 15 trials in each period. It can be seen that over the entire period of the experiment the same quantity of milk accumulated in the udder before the morning milking. The interval before the morning milking was 10 hours. These two factors, which determine the reaction of the cows to milking, remained unchanged and could not therefore affect the results of the experiments. The surroundings where the cows were milked during the project were also the same. The cows were maintained in the same section of the buildings and the specified order of daily routine was strictly observed. The buildings were maintained at a temperature around 10°C.

TABLE 41. TOTAL NUTRITIVE VALUE OF FEEDING SUPPLEMENTS

Period of experiment	Kind of feed	Quantity of feed (kg)	Nutritional units (kg)	Digestible protein (g)
March	Coarseground Barley Sunflower seedcake	1.2 0.5	1.98	294
April	Corn silage	9.0	1.80	126
May	Lucerne hay	4.0	1.96	464

Leaving the cows for more than 13 hours in the second period of the trial with no feed and without supplementary feeding led to a fall in the completeness of milk letdown by 21.8%. Supplementing with concentrates in the third period of the experiment enhanced milk letdown by 10.2%, but this still was lower than in the first or control period by 11.6%. Supplement of silage in the fourth experimental period stimulated the milk letdown reflex as a result of which the milk collection was raised by 5.9% in comparison with control (first period) and by as much as 27.7% compared with the second period. Similar results were obtained when the cows were fed with supplements of lucerne hay in the fifth experimental period.

How then do we explain the results of these experiments? Why is it that with the same allowance but with a 13-hour break when the animals are not fed the concentrates the milk letdown after this "fast" is so severely inhibited? What is the reason for the fact that feeding the cow with concentrates three hours before milking raises, but does not fully restore, the

TABLE 42. INFLUENCE OF FEEDING REGIME OF COMPLETENESS OF MILK LETDOWN IN COWS

Periods of experiment	Quantity of milk collected (liters)	Residual milk (liters)	Total milk in udder (liters)	Completeness of milk letdown (%)
January	10.0	1.5	11.5	87.0
February	7.3	3.9	11.2	65.2
March	8.0	2.6	10.6	75.4
April	10.5	0.8	11.3	92.9
May	9.9	0.9	10.8	91.7

index of completeness of milk letdown to what it was when the cow was being given her usual ration (first period of the experiment)? Finally, why does supplementary feeding of the cows three hours before milking with silage or lucerne hay give higher values of completeness of milk letdown than with the usual regime of feeding? It will be seen from Table 41 that the improvement in the milk letdown indices in the animals with supplementary feeding is not related to the nutritive value of the supplements. They are rather related to the influence exerted by the bulk of the ingested feed, i.e. to the mechanical effect of the food mass on the receptors of the pregastric chambers of the stomach. Studies by Sineshchekov [163] showed that in cattle the food mass passes through the gastric chambers and enters the intestine in the course of 7–10 hours. Therefore it can be assumed that in the present experiment with cows the pregastric chambers (rumen, reticulum and omasum) and the stomach (abomasum) were essentially free of chyme at the time of milking. As Salmin [152] showed, the state of satiation or feeling of hunger in ruminants depends on the degree of filling of the gastro-intestinal tract and the nature of the food. Based on Sineshchekov's [162] and Grachev's [72] work and on Fedii's [62] research on cows with fistulas, cited by Sineshchekov [161], we may assume that the state of the digestive tract and particularly of the pregastric chambers of the stomach, the stomach and the small intestine may exert a direct reflex effect on the function of the cow's mammary gland. This allows us to assume that in our experiments with the emptying of the pregastric chambers of the stomach by milking time a feeling of hunger arose in the cow. This was transmitted to the cerebrum and from here signals were conveyed to the mammary gland, causing inhibition of milk letdown. Just as an increase in the volume of residual milk in the udder leads to a decrease in the rate of milk secretion, the posited influence of the pregastric chambers on the completeness of milk letdown may be considered as a protective reaction by the cow to prevent diversion of nutritive substance from the organism due to disruption of her regular intake of feed. However better conceived experiments are required before this assumption can be either accepted or rejected. But it is quite probable that owing to insufficiency of bulk in concentrate feeding the feeling of satiation was not produced in the animals and hence extent of completeness of milk letdown was lower than under the normal regime of feeding. On the other hand, supplementary feeding with silage and hay was conducive to a better filling of the pregastric chambers with food than with the usual feeding regime. This produced a feeling of satiation in the cow and the receptors of the pregastric chambers were stimulated. This feeling was conveyed to the brain from where signals were transmitted to the mammary glands. Even the milk letdown was not stimulated by this at least it was not inhibited. The outcome is that the completeness of stripping of milk from the udder is higher with supplements of silage and hay than with the usual feeding regime. What practical conclusion can be arrived at from these studies? Since the milk letdown reflex functions more effectively in cows when the pregastric chambers are replete with food the distribution of the daily allowance to the animals should be such that a feeling of satiation is produced in them before each milking. There should be an interval between feeding and milking so that the feeding reflexes fade out during this period and the animals become tranquil in preparation for milking.

Maintenance regime of cows

The readiness of cows to be milked depends on whether they are allowed the normal period of rest. Cows should have a rest period of not less than 9–10 hours in the day. It is very important to provide normal conditions for the animals to lie down during the night, particularly during loosehouse maintenance. The regime and the number of feedings and milkings should be so adjusted that the animals are allowed plenty of time for rest in the day and for lying down at night. It is necessary to have buildings and dry litter for the cows in loosehouse maintenance during hot summer days and rainy days.

To prepare the cows properly for machine milking it is very important to know the effect of low winter temperatures on the condition of the animals and the magnitude of the milk collection. It is thought that the maintenance of cows in buildings at temperatures below zero has no negative effect on their health. As the atmosphere cools down there is an increase in the heat output of the animal and in consequence of this the consumption of feed and production of heat also increase. Due to this the feed requirement increases or if feeding is limited there is a decrease in milk yield. Loosehouse maintenance of the cows is progressive technology and is partly linked with the technology of continuous machine milking of the animals. But this does not mean that all that has been learned in dairy farming from experience over many centuries should be discarded right away. On the contrary, the methods of maintenance and handling of cows developed for housed maintenance [80] are fully applicable to loosehouse maintenance.

Technology of continuous milking of cows in milking installations

At present milking parlors of the herringbone, tandem and UDM–8 types are usually built for herds of 150–250 cows while installations of the rotary type are built for herds of 300–600 cows. This results in the milking installation being idle for the greater part of the time while the working day of the milkmaids, cattlemen and other farm workers is split up as on a farm where the cows are milked in stalls. Under these conditions the milkmaid has to report to the farm by 4-5 in the morning and stay until 8-9 in the evening, using the intervals between milking for recreation. In this scheme of working the introduction of milking installations is unable to approximate the work on the dairy farm to industrial work, i.e. to make the working day continuous. Besides this, the low utilization of the milking plant and consequent idle time leads to huge unproductive expenditure in capital inputs by the dairy farming organization which evidently cannot make machine milking economically effective by and large. The study of the biological basis of the technology of continuous milking of cows in milking systems and the elaboration of this technology are of great national economic importance. The essence of the continuous machine milking technology consists in milking out groups of cows according to a staggered schedule. It does not make any difference to the cow whether she is milked on a 4 a.m. to 4 p.m. or a 6 a.m. to 6 p.m. schedule but it is important that the milking schedule be strictly adhered to. The technology of continuous milking makes it possible to increase to a considerable extent the load on milking system, which should work for a minimum of 14 and a maximum of 21 hours a day. This would permit the introduction of a continuous 7-hour working shift for milkmaids, cattlemen and other workers on the dairy farm and at the same time reduce the capital outlay per cow. With continuous machine milking the load on the tandem, UDM–8 and UDS–3 plants may attain 400–500 cows, on the herringbone plant 600–700 cows and on the rotary type of plant 800–1,500 cows. The technology of continuous milking excludes idle time for milking equipment, enhances the effectiveness of utilization and permits concentration of a large number of cows at one place for the process for milking. With continuous machine milking it becomes possible to concentrate on milk production specialization and enhancement of the productivity of labor of the workers in dairy farming.

Time of milking and size of herd

Since conditioned reflexes are generated at milking time and are firmly linked to it, in the case of loosehouse cows and continuous milking it is necessary to have a situation where the time of each milking is strictly adhered to. How can this be brought about in practice? In milking installations the cows cannot be milked in the same order as, for example, when they are milked in stalls. Of course, it is a matter of observation that among the animals of a herd a certain hierarchy is established. In a milking installation of the box type of building as a rule the strongest animals enter first, then the less strong and lastly the weakest. But this order is not always maintained, and in large herds of 150–200 cows deviation from the usual milking time may be 1–3 hours. Owing to this, intervals between milkings may vary (being either longer or shorter) by 2–6 hours. This variation in the intervals between individual milkings is highly undesirable since under such conditions reflexes are disrupted at milking time and this practically rules out the possibility of effective adjustment of the animals for milking. This inhibits milk letdown and leads to a reduction in milk yield.

In the technology of continuous milking the strict observation of each milking time depends on the size of the herd and the capacity of the milking system. The time taken to milk a herd of cows in a milking installation should not exceed 40–60 minutes. In rotary types of installation the number of cows in a herd should not exceed 100. In the herringbone type, where the capacity is 60–70 cows per hour, the number of milking cows in the herd should not exceed 60–70 head. In the case of parlors of the tandem, UDM–8 and UDS–3 types this figure would be still lower, 50–55 cows. The size of the sections of the cow barn depends on the type of milking system. Only with the maintenance of this requirement is it possible in continuous milking technology to keep strictly to the fixed time sequence needed for the cows to adjust. Unfortunately this requirement is rarely fulfilled when a new milking system is adopted. In the stall milking system the size of the herd attains 100–200 cows. This size of herd is automatically preserved when the cows are transferred to mechanical milking without consideration for the type of milking installation, its capacity and the requirements relating to maintenance of the routine of work and the milking regime. When adopting the technology of continuous machine milking the size of the herd should be determined on the basis of the specific conditions that would ensure the milking out of all the cows in one herd in the course of 40–60 minutes.

Adjusting the cows to milking

It is known to every stock breeder that in order to bring about normal setting before milking the cow should be brought from the pasture to the milking place half an hour before milking so as to allow the animal to settle down. Some of the animals lie down and peacefully chew the cud while others stand around waiting for their turn to be milked. The state of satiation after grazing, the accustomed place and time of milking, the tranquillity, the uniformity of the milking stimuli, the absence of any agitation on the part of other animals—these factors constitute the environment in which the adjustment of the cow to milking takes place under conditions of pasture maintenance of the animals.

A roughly similar environment should be created in the reception area of a milking installation. It should be remembered that the adjustment of cows to milking is a very necessary part of continuous milking technology. Without this preparation it is not possible to milk the cows effectively by milking machine in milking parlors. Therefore the reception area, its layout, the color of the walls, lighting, design of the floor, system of drainage and other features should all contribute to the creation of a relative degree of comfort to which the cows can become habituated. Unfortunately, in the practice of machine milking of cows, these aspects of regular preparation and adjustment of the animals to milking with the aid of natural factors which constitute their environment in the reception area before milking do not receive sufficient attention. Consequently the effective functioning of the milking machines is severely handicapped.

Temperature conditions of the reception area

This factor exerts an influence on milk letdown only in winter. But in spite of this it should be taken note of so that a decrease in milk productivity or premature drying up, or mastitis does not set in. The atmospheric temperature in winter in the reception areas should be the same as in the milking installation, not lower than 15°C. For heating this area it is advisable to use a system of radiators. A higher temperature in the reception areas serves to maintain a steady temperature regime in the milking rooms, eliminates the possibility of drafts and creates the best conditions for the servicing staff. But the chief importance of heating the reception areas lies in the following: In continuous milking technology and loosehouse maintenance (in conditions where the reception areas are not heated) the cows do not enter the milking installation from a heated cowshed but from the pasture. The skin of the teats and udder gets cold in freezing weather and the nerve endings of the teats become insensitive and do not respond to the stimuli of washing the udder and slipping of the teat cups over the teats. Due to this, the milk letdown reflex is weak and the cows do not yield enough milk to the milking apparatus. This is of particular importance in milking cows in the rotary type of installation. In the absence of reception areas for cows the animals enter the rotary installation directly from outdoors. Within the next 1–2 minutes they are subjected to treatment: the udder is washed and massaged and the teat cups are put on. In another 6–8 minutes they have left the milking stall. The teats of the cows cannot be warmed up thoroughly in that short time and the milk letdown reflex will be weak and milking effectiveness low. In such conditions the cows lower their yield sharply and the working of the milking installation is rendered uneconomical. In the herringbone type the entry route into the milking stalls is a little different. When a group of cows fill up one side of the herringbone they are not immediately milked but only after 6–7 minutes, when the group of cows on the other side have been milked out. But this lapse of time is not enough for the recovery of the normal sensitivity of teat nerve endings and so milk letdown is not sufficiently active.

After the cows are milked they should not be sent out into the cold at once. The animals should be kept for a while in a departure hall so that the udder dries off and cools down. In this way they are protected from mastitis. It may be mentioned here that though the low temperature at the range and in the cowsheds in loosehouse maintenance requires additional expenditure of energy to warm up the body it does not harm the health of the animal. However, for successful machine milking of cows in winter conditions it is necessary that the milking hall and reception areas be kept at above-zero temperature levels. In the case of large farms where the technology of continuous machine milking and loosehouse maintenance of cows have been introduced as, for example, on the mechanized farm in the experimental establishment "Kutuzovka" of the Scientific Research Institute of Animal Husbandry of Forest Steppe and Polessie* of the USSR [132], the study of optimum temperature conditions for milk letdown in cows is of considerable practical importance.

Physiological requirements of the technology of machine stripping of milk from the udder

Since milk letdown in cows is of brief duration and cannot be repeated it is very important that all the operations connected with milking out the udder should not be interrupted. All operations such as washing, drying, preliminary massage, putting on the teat cups, milking, final massage, machine stripping, pulling off the teat cups from the udder and inspection of the condition of the udder after milking—all should follow one another uninterruptedly. These operations are performed quickly and energetically but without being unnecessarily rough and hasty. A thorough observance of these requirements makes it possible to milk cows fully in milking

*The vast alluvial plains—General Editor.

parlors. In the herringbone system with twice-a-day milking the following order of preparation of the animals for milking is recommended. After washing the udder of the first cow the udder should be wiped at once, then the teat cups put on the teats. If the milking machine withdraws the milk vigorously the milkmaid should set about her task for the preparation of the second cow, then the third and so on. If the milk does not enter the milk pipe the udder is massaged until the cow lets down her milk and only then does the milkmaid attend to the next animal. After the teat cups are put on the teats of the last cow, the eighth, in the UDE–16 installation, and the cow lets down her milk the milkmaid goes to the first cow, attends to her final massage and machine stripping, then pulls off the teat cups and checks the state of the udder. These operations are repeated with the rest of the cows.

Often, in installations of the rotary type a special stand is constructed some 2–3 meters away from the rotary platform for washing the udder. One of the recommendations on machine milking of cows [2] suggests that the udder be prepared for machine milking in rotary type installations at a separate stand. It is true that in this method the washing of the udder stimulates milk letdown, but the cow has to wait for a minute or two until her turn comes to approach the milking stand. After the onset of milk letdown the cow has to move and align herself properly in the milking stand on the rotating platform. On her way to the stand the food reflex is stimulated by the offer of concentrates. This system of preparing the cow for milking requires the animal to carry out three physiological functions of widely different nature and purpose simultaneously: locomotion, in which the central locomotor coordination of the brain is involved: reflex milk letdown, whose fulfillment requires nonparticipation of the organism in other actions; and the reflex feeding which involves consuming her ration of concentrates. What does all this lead to? Milk letdown in the cow is a reflex process which is easily inhibited. The locomotion reflex is more stable, especially where, as in this case, its accomplishment is connected with and subordinate to the feeding reflex. Thus the function of locomotion in combination with the feeding process diverts the cow's attention and suppresses the reflex of milk letdown. This leads to incomplete milking of the cow in the milking parlor. In rotary milking parlors all operations connected with preparing the cow for milking should be carried out in the milking stand on the revolving platform after the animal has assumed a steady position and quieted down. Irrespective of the design of system offered by inventors for milking cows its physiological basis in the future should be the uninterrupted flow of operations concerning the milking process. Without this precondition the normal onset of the milk letdown reflex and effective evacuation of milk from the mammary gland will not be possible in machine milking.

Feeding with concentrates

In view of the fact that the milk letdown reflex is easily inhibited by various factors that distract the cow's attention it may be assumed that other linked reactions appearing simultaneously with milk letdown may influence it. Then how do we explain the combination of milking with feeding concentrates in the milking parlor? Data were obtained in Sineshchekov's laboratory [53] that showed that when the cow is being milked the process of digestion is somewhat inhibited. In view of these data it is justifiable to assume that to some extent feeding the cow at milking time should also have an influence on milk letdown activity. However, data obtained in our experiments show that supplementary feeding of cows with concentrates at milking time in the milking parlor may yield high values of milking by machine. These data indicate the possibility of combining a highly active milk letdown reflex with the feeding reflex. However, the conditions under which this combination is possible are not yet known.

Experimental analysis of these conditions is very important because in conducting machine milking it is necessary to investigate how along with supplementary feeding of concentrates one can maintain the preparatory adjustment to milking in the cow and ensure a high degree of milk letdown reflex activity. On the basis of general physiological conclusions it may be presumed that for the appearance of the milk letdown reflex, as for any other reflex, the ideal conditions would be those where all other reflexes are either suppressed or subordinated to the reflex whose fulfillment is predominant at the moment. In view of what has been said it is necessary to organize the supplementary feeding of cows with concentrates in milking parlors in such a way that feeding reactions do not inhibit the milk letdown reflex. In herringbone installations the single feed ration should be so calculated that as the cows are being milked on one side the cows on the other side are eating up their concentrate and getting set to be milked. But if this cannot be achieved it is necessary to organize supplementary feeding so that milking begins at least after the cows have eaten half their allotted feed. At this time the feeding reaction is slack and the effectiveness of the milking stimuli on the udder will be higher.

In milking parlors with stalls of the walk-through type like the UDM-8 and UDS-3 work should be organized in the following manner: The milkmaid coaxes the cow into the milking stall and lets her go to the feed box. After this she prepares to milk the next cow, which entered earlier and will have eaten part of her feed by this time. In rotary milking parlors the feed should be dropped into the feed box from the feed bin before the cow approaches the milking stall. In this kind of organization of supplementary feeding the cow soon sets about eating the concentrate on her way to the milking stand as she eats it moves about to take up a convenient position in the stand on the rotating platform. The preparation of the cow for milking under these conditions is better begun a minute or so after she enters the milking stall. By this time the cow will have eaten part of her feed concentrate and taken up a comfortable position in the stall so that stimulation of the udder will be quite effective. In milking installations the beginning of supplementary feeding of cows when their feeding reflexes are being strongly expressed should not be combined with the preparation of the cow for milking. The feeding reflexes at this time may inhibit the milk letdown reflex. In order to make the milking stimuli more effective the preparation of the cow for milking should be carried out after the cow has eaten part of her ration of the feed and the feeding reaction has consequently slackened.

Supplementary feeding of the animals with concentrates is a very important element in the technology of milking cows in milking parlors. In the practice of machine milking it is very necessary to determine the place of supplementary feeding in relation to the schedule of the milking parlor and to time it properly to study the effect of the supplement on the completeness of milking out of the udder and to prepare on the basis of these data a technological scheme for preparing the cow for milking.

Preparation of the cow for milking

In preparing the cow for milking it is necessary to meet a number of requirements on which the milk letdown reflex depends. The temperature of the water for washing the udder should not be lower than 45–50°C. It is necessary to maintain this temperature in winter as well as in summer. The teat cups should be warmed up in water at 60°C before they are put on the teats. In the practice of machine milking the working out of an automatic arrangement to maintain the teat cup liners at a constant temperature of 40–45°C throughout the period of milking would be of great value. The milkmaid also should warm her hands in hot water before she begins the task of preparing the cow for machine milking. Special studies have shown [36] that when the cows are milked with hands whose skin temperature is 22–28°C the udder temperature falls but when the hands were warmed in hot water before milking and milking was done with a hand temperature of 32–34°C, the temperature of the skin of the udder rose immediately. These data show that massage of the udder and milking the cow with cold hands cause a negative reaction in the animal.

Washing the udder is usually done with the help of a bucket. But often the parlor is equipped with a shower basin. This consists of a metal funnel with a nozzle connected to a hose. Warm water issues from the hose. The base of the funnel consists of a metal plate with a small hole. The water is turned on by pressing a lever or opening a stop cock. With this arrangement the force of the jet of water depends on the pressure at which the water issues from the tap. Observations show that washing the udder with a strong jet startles the cow and this may lead to inhibition of milk letdown. The jet from the shower wash should be small and should not depend on the pressure of water in the pipe. The basin should be fitted with a regulator to let the water into the shower wash. After washing, the udder is immediately wiped with a dry towel and massaged. Washing of the udder should take 8–10 seconds and drying not more than 10 seconds. This constitutes the so-called stimulation of milk letdown in cows before the teat cups are put on the teats. The well known American worker in the field of lactation and machine milking, Petersen, attaches great importance to stimulation of milk letdown in cows before the teat cups are slipped onto the teat [143]. However, special tests conducted in England [48, 50] and in the USA [94, 166, 195] on the influence of udder washing and massage over periods of 15 seconds to 20 minutes before putting on the teat cups gave contrary results. Whittlestone [198, 199] as well as some Australian workers [131] recommend massage of the udder for 30–60 seconds before putting on the teat cups. According to their data this raises the rate of milking and the yield of milk. However, these methods of stimulation of milk letdown have not been supported by experimental testing on the basis of residual milk.

With twice-daily milking, when the udder is fairly full of milk and the milk letdown reflex is highly effective, the period of preparation of the cow for milking should not exceed 30 seconds, else it may become necessary to prepare the cow for thrice-daily milking. On the majority of farms the milking intervals are 10, 8 and 6 hours. For the morning and noon milkings the cow is prepared for milking in the same way as for twice-daily milking. At the evening milking the udder is not quite replete with milk and the teat cisterns are mostly empty at the start of milking. The impact of the vacuum on an empty teat, particularly in milking with 2–phase milking machines, may cause bleeding and this may lead to mastitis. To avoid this at the evening milking the udder is energetically massaged for a minute after washing and drying so that the teats may fill up with milk. After the cow lets down the milk and the teats become taut, the massage of the udder is discontinued and the teat cups are put on the teats.

Stripping milk from the udder

The teat cups are slipped over the teats in the following way. After opening the vacuum tap of the rubber hose conveying the suction force from the pulsator to the collector and teat cups the teat cups are held in the right hand and their rubber delivery vacuum nozzles are pressed onto the collector. Then with the left hand a teat cup is slipped over the left rear teat, and the other teat cups successively over the right rear teat, left fore teat and the right fore teat. It is necessary to observe the following precautions: After the teat cup begins to grip the teat it should be slightly drawn out so that severe clamping of the cup to the teat is avoided. That would make the cow uneasy and inhibit letdown of milk. Putting on the teat cups too quickly may lead to folding up of the teat with the tip upward so that the quarter remains unmilked. The teat cup should be fixed on the teats in the suction phase. This avoids turning up of the teats. With proper preparation of the cow for milking the milk will begin to flow out of the teats vigorously some 3–5 seconds after the teat cups are put on. The duration of milking is on an average 4–5 minutes, sometimes as little as 2–3 minutes. The milking apparatus should not be allowed to remain on the udder after the milk has been stripped and the teats appear to be empty. In slow-milking animals this may cause mastitis and in fast milking animals hyperemia of the teats. As a result the cow may develop a negative reaction to machine milking, leading to retention of milk at subsequent milkings. Hence milkmaids and specialists who think that the longer the milking apparatus is allowed to hold onto the teats the greater the quantity of milk stripped are wrong. After the flow of milk from the teats slows down or stops the teat cups are pulled downward and forward and held for 10 seconds in that position (machine stripping). If no milk comes the teat cups are pulled away. The teats and quarters of the udder are examined and the udder is wiped with a dry towel, completing the milking process. The removal of the teat cups from the teats may be made automatic. A mechanism has been designed to cut off the vacuum in the teat cups at the close of milking [164, 193]. In the USA the Perfection company has designed a milking machine incorporating automatic disconnection of teats that have been milked out. It is to be regretted that such devices have not come into wide use. The effect of this form of removal of teat cups on completion of milking has not yet been studied though the prospects for use of such milking devices are obvious. Some specialists advise a final massage of the udder at the end of milking and before the teat cups are pulled off. According to them this should stimulate the letdown of milk. But reflex milk letdown is either weak or will have stopped toward the end of milking. Therefore the effectiveness of udder massage on milk letdown under these conditions will be weak.

Hand stripping

Proper termination of machine milking is as important as the preparation of the cow for milking. How this has been done governs the further formation of milk, the volume of milk collection at the next milking and the state of the cow's udder. In view of this a few specialists attach great importance to manual stripping after the teat cups have been pulled off. They believe this facilitates a more complete stripping of milk from the udder. It is supposed that with the help of hand milking after the teat cups have been pulled off the udder it is possible to induce a second letdown of milk, a repeated milk letdown. This would compensate inadequate action of the milking apparatus on the cow's udder. Hand stripping of cows is a laborious business. It requires much time and, besides, complicates the technological process of machine milking in the milking parlor. According to average data from physiological research on milking installations the average duration of milking by machine is 4 min 8 sec, and the average duration of hand stripping is 1 min 24 sec. Thus hand stripping requires an additional outlay of 30% of the time spent in milking the cow by machine. It is necessary to consider whether this recommendation to hand strip the cow after machine milking is sound.

In the UDM–1 type of parlor, in response to hand stripping after the teat cups were removed some 5% of the total quantity of milk accumulated in the udder at the start of milking and 9% of the fat were collected. After this hand stripping there remained in the udder some 14% residual milk and 33% fat. These data confirm that a second milk letdown does not follow hand stripping in cows. From this it is evident that the use of hand stripping along with machine milking of the cow is not physiologically sound. From the economic point of view also hand stripping is not profitable since much time has to be spent on this operation while the return in milk and fat is small. But it is not possible that if the milk that was hand stripped had been left there it would have had a negative effect on the condition of the udder and the subsequent secretion of milk? Leaving 0.36 liter of milk in the udder is not going to have much influence on the rate of milk secretion, though in order to raise the rate of milk secretion it would be necessary to withdraw from the udder not only the hand-stripped milk but also the residual milk (0.99 liter). For more complete stripping of milk from the udder in machine milking it would be better to employ not hand stripping, whose efficiency is low, but udder massage after the teat cups have been placed in position on the teats. In the UDM–8, in response to massage an additional 10% milk and 21% fat were drawn out of the udder, i.e. twice as much as by hand stripping. Udder massage does not prolong the time spent on milking but hand stripping does, appreciably.

Hand stripping of cows should not be employed along with machine milking for another reason. Cows get accustomed to stimulation of the teats produced either by hand or by machine. In animals a fixed connection is developed between the nature of the stimulus and the strength of the reflex action evoked by it. But the two stimuli—the teat cup and the milkmaid's hand—are dissimilar. At a single milking the cow cannot react in the same way to the influence of stimuli so different in character in quick succession. This may lead to derangement of the unconditioned milk letdown reflex either by the action of the milking machine or by hand milking. This is at the root of the fact that cows frequently transferred from hand milking to machine milking or vice versa, or regularly hand-stripped after machine milking, do not milk out well either with the milking machine or by hand milking.

In adopting a milking system a number of problems face the practical workers of the dairy farm. The chief problems that interest animal husbandry workers are what measures are to be undertaken for the milk yield of the cows not to decrease with the introduction of machine milking? What are the characters by which cows are to be selected for machine milking? How are the cows reconciled to the milking parlor without the use of force? How are cows switched from hand to machine milking [191]? How does one overcome delayed milk letdown and ward off mastitis in machine milking? and so on. None of these questions has been adequately researched. We have to bear in mind that machine milking technology is overloaded with a surfeit of details or trivialities. The relevance of each of them has been only partially investigated so that to an inexperienced reader all this may appear inessential work which does not merit attention. But it is on these "trivialities" that the activity of the milk letdown reflex of the cow in the milking parlor, the completeness of the stripping of milk from the udder at each milking and, consequently, the economic result of the operation depend.

Machine milking with stall maintenance of the cows

Machine milking in milking parlors is most effective with loosehouse maintenance of the animals. But since this style of maintenance is not yet widespread among the dairy farms of kolkhozes and sovkhozes it is necessary to summarize the main features of the technology and organization of machine milking in stall maintenance of cows. Even when the cows are maintained in the tie-up stalls they can be milked in the milking parlor. For this purpose chains of 10–25 cows are established in the cowshed. Milking is organized in the following way: In the center of cowshed a small pre-milking area is set up. It is divided into two sections. From the first section the group of cows is guided along a passage way to the milking parlor, which directly adjoins the cowshed. After milking the cows enter the second section and from here they are guided back to the stalls and tied up. This kind of milking technology for cows may also be recommended for the combined maintenance of cows [130] that has been followed in Ukraine in recent years. In the combined maintenance system the cows are maintained in free condition in summer and are tied up in stalls in winter.

In recent years the milking of cows at pipeline milking installations has spread widely. For 100 or 200 cows a transparent milk pipeline of unbreakable glass or plexiglass is set up for this purpose in the cowshed. The pipeline extends all along the stall at a height of 170–180 cm. The cows are milked out through a milk collector which is suspended from the milk pipeline or directly into the pipeline. Portable milking machines are connected to the milk pipeline. Trials have shown that by employing this system in tie-up cowsheds, equipped with transporters for distribution of feed and removal of dung, it is possible for one milkmaid to milk out and look after 40–50 cows. For washing the udder of the cows with warm water special mobile cisterns may be provided or hot water may be supplied by pipe from an electric boiler. Machine milking in stalls is more labor-consuming and less convenient work than in milking parlors with loosehoused cows. To the existing deficiencies of this mode of milking may be added the fact that in milking the cows in tie-up stalls it is not possible to organize a continuous working day for milkmaids. Milkmaids and cattlemen are obliged to come to the farm by 5 o'clock in the morning and stay until 9 o'clock at night with rest periods between milkings.

The problem of hard udder in machine milking

There is no definition of the term "hardening of the udder" in physiological literature. The part played by hardening of the udder in an outbreak of mastitis is not known. Zaks [205] has studied the phenomenon of hardening of the mammary gland in veterinary practice. According to him the mammary gland goes hard due to the retention of milk in consequence of weak activity of the milk letdown reflex.

In view of the fact that in machine milking incomplete stripping of milk from the udder is the principal cause of declining productivity, the causes of hard udder were investigated in detail at the Donetz agricultural experimental station. The condition of hardness was defined and measures to combat this phenomenon were worked out. Lumpiness is a specific condition of the udder or sometimes of particular udder quarters, which cannot be termed a pathological state. The distinctive features of hard udder in cows are as follows: One or more quarters of the udder are "stony" to the touch. The condition is not painful. There is no increase in the body temperature of the animal, her appetite is normal, chewing the cud is not disrupted. The hardening of the udder is due to disruption of the milk letdown reflex, incomplete stripping of milk from the udder quarters, frequent upturning of the teats with the sphincters pointing upward because of faulty application of the teat cups on the teats, increased sensitivity to vacuum, irregular development of the quarters of the udder and so on. The hardening of the quarters of the udder leads in the majority of cases to mastitis. Observations have shown that of every 10 cases of hardened udder six to eight cases terminate in mastitis unless the animals are attended to in time. In view of this, hard udder may be considered a preliminary stage of bovine mastitis. Measures against hard udder should be an integral part of machine milking technology in milking parlors.

Hardening of the quarters of the udder is dealt with in the following way: During the examination of the condition of the quarters of the udder before the teat cups are removed attention is paid to the condition of the quarters, whether they are empty, soft, full or, lastly, hard or "stony" to the touch. Where one or more quarters are "stony" to the touch it is considered hardening. After one or two quarters of the udder are found hard the teat cups are removed. First of all an attempt is made to milk any quarter that is hard by hand. But other quarters of normal consistency are not milked by hand. If some milk can be withdrawn from the cistern of the quarter and it is emptied then part of the milk from the alveolar spaces passes over to the cistern and the "stony" condition of the quarter is corrected: it now becomes soft to the touch. Observations have shown that in these cases a repeated letdown of milk or milk letdown reflex does not, as a rule, take place but the teat sphincter muscles and also the smooth muscle fibers of the cistern of the quarter relax under the influence of indirect milking stimuli. As a result of this there is the possibility of withdrawing cistern milk from the quarter. When the consistency of the quarter becomes soft to the touch due to hand milking this is stopped and the cow is sent back to the herd. Often, however, hardening mostly of the alveolar part of the quarter is found. In such cases hand stripping of the quarter does not return it to the normal state since there is no second milk ejection. The quarter should not be allowed to remain in this hard state until the next milking. The cow should be led out of the milking stand and 20 IU of oxytocin injected into her jugular vein. This is followed by hand stripping the milk from the udder. When all the quarters are found to be hard hand stripping should not be employed. Oxytocin must be injected, the udder massaged and the teat cups slipped on again. Milk ejection takes place because of the effect of oxytocin and the milking machine will withdraw this milk. The udder will again be soft to the touch. Trials showed that with the help of repeated machine milking some 6–10 liters milk is withdrawn in addition to the 1–5 liters milked out at the principal milking. The quantity of milk withdrawn from the udder in two milkings of a hardened udder considerably exceeds the maximal indices of single milk collection. These data indicate that hardening of the quarters appears as a result of severe overfilling with milk of the cavities of the milk gland. It is quite possible that the smooth muscular tissue is stretched to the limit with this serious overfilling of the udder. Its capacity for normal functioning is partially or fully lost and the milk letdown reflex is not produced. The method of dealing with udder hardening as set down above allows full protection against mastitis, which is the result of such hardening. Loss of milk is prevented and lactation, which would have been disrupted if udder hardening had developed into mastitis, is reestablished. The use of these methods to prevent hardening of the udder in machine milking does not require ambulatory maintenance of the animals. After injection of oxytocin the cow may be milked by milking machine at the next milking. In such cases the udder has to be massaged in order to cause a satisfactory inflow of milk.

Udder evaluation in selecting cows for machine milking

It is thought that machine milking does not impede increasing the milk yield of cows up to 6,000 kg of milk per lactation [124], though to obtain 10,000 liters of milk per lactation hand milking is still recommended [108]. Special trials [81] showed that the milk yield of first-calf cows with machine and hand milking did not differ much. It amounted to 5,253 and 5,702 kg respectively per lactation. It is known, however, that with loosehouse maintenance and continuous flow technology in milking installations it is possible to obtain higher annual milk yields. In the USA, for example, a farm has been designed and established for 1,100 cows on these lines [133]. The annual average yield of milk per cow on this farm is 5,400–5,900 kg. The farm employs 22 people. Large farms of 1,000 cows or more employing continuous milking technology in their milking parlors with the productivity of the animals exceeding 4,000 kg per lactation are to be found in Italy and other countries. All this shows that continuous milking technology in milking installations need not be considered as impeding high productivity of cows. The selection of cows to be milked by machine is a factor of great importance. It has been observed that in the case of first-calf cows the speed of milk letdown with machine milking does not depend on the length and diameter of the teats but on the dimensions of the teat aperture [84]. It may, however, be borne in mind that teat length and diameter are important morphological indices of the udder. Cows whose teats are too long (10–12 cm), or too short (2–4 cm), or too stout or too thin are not suitable for machine milking. Since there is a link between the milking speeds of mother and daughter [51] there is a proposal to classify [77] and select cows on the basis of this character [148, 200]. The first results of this work [160] show that cows with high milking speed gave a milk yield of 4.5–5.0 thousand kg milk per lactation compared to 3.8–4.8 thousand kg milk given by slow milkers. It may be presumed that the higher milk yield at high milking speeds is linked above all to more complete stripping of milk from the udder.

The ideal udder for machine milking is one with uniform development. The shape of the udder has an influence on the proportional development of the quarters. An animal with a bowl-shaped udder is to be desired most since the quarters are all equally developed. Of course, Martyugin [123, 125] believes that teat cups hold on for a shorter time on less productive quarters in comparison with more productive ones. But Cherkashchenko [39] showed that in cows with bowl-shaped udders the retention of teat cups on the less well developed front quarters was 30 sec longer while in cows with caprine (goat-like) udders the difference was 2 min 24 sec. This amounts to 46.3% of the duration of milking of the front quarters. These data show that milking out cows having uniformly developed quarters requires less time than in the case of cows having nonuniformly developed quarters.

TABLE 43. MILK LETDOWN AND DEVELOPMENT OF QUARTERS OF THE UDDER IN THE COW TSYGANKA WITH THE FRONT QUARTERS MILKED IN THE FIRST INSTANCE (MEAN DATA OF SIX EXPERIMENTS)

Indices	Udder quarters				Total udder
	Right front	Left front	Right rear	Left rear	
Yield of milk (ml)	2,140	2,090	2,145	2,165	8,540
–Do– as % of milk from the udder	25.0	24.5	25.0	25.5	100
Residual milk (ml)	210	180	975	895	2,260
Total milk (ml)	2,350	2,270	3,120	3,060	10,800
–Do– as % of total milk from the udder	21.5	20.6	28.1	29.8	100

Recently Gar'kavyi [64] published data on the evaluation of bulls in respect of the udders of their daughters. For a point scale evaluation of the udder this author elaborated a procedure in which the requirements of the cow's udder were considered from the point of view of machine milking. This procedure was taken as the basis of Instructions on the Evaluation of the Udder [3], which was issued by the Ministry of Agriculture, USSR. The chief factors determining the suitability of cows for machine milking are the shape and size of the teats, completeness and speed of milk letdown, development of udder quarters and shape of the udder. Development of the quarters is at present determined on the basis of the quantity of milk yielded by each quarter when milked separately into a four-chambered milk pail or by a four-chambered milking apparatus [41, 95, 104, 151]. However, the magnitude of the yield of the quarters cannot be considered a

TABLE 44. MILK LETDOWN AND DEVELOPMENT OF QUARTERS OF THE UDDER IN THE COW TSYGANKA WITH THE FRONT QUARTERS MILKED IN THE FIRST INSTANCE (MEAN DATA OF SIX EXPERIMENTS)

Indices	Udder quarters				Total udder
	Right front	Left front	Right rear	Left rear	
Yield of milk (ml)	1,300	1,160	3,100	3,050	8,610
–Do– as % of milk from the udder	15.0	13.5	36.0	35.5	100
Residual milk (ml)	980	870	320	280	2,450
Total milk (ml)	2,280	2,030	3,420	3,330	11,060
–Do– as % of total milk from the udder	21.9	19.0	30.0	29.1	100

sufficiently precise criterion for evaluating the development of a quarter. In view of this we measured the development of the quarters on the basis of our own procedure. The cows were milked out in the morning into a four-chambered milk pail. In the first series of experiments the front quarters of the udder were milked out first and in the second series the rear quarters first. The completeness of milk letdown was determined once in three days. In Tables 43 and 44 we give the results of these tests with the cow Tsyganka. In cases where the front quarters were milked out first more milk was drawn out from their alveolar spaces and less from the alveolar spaces of the rear quarters. The development of the quarters considered on the basis of the indices of milk yield appeared in this case uniform. But the inclusion of the residual milk of each quarter and estimation of the development on the basis of total milk accumulated in the quarters at the start of milking showed that the rear quarters of this cow were better developed. Conversely, a sharp disproportion could be seen in the development of the quarters estimated on the basis of milk yield when the front quarters were milked out after the rear quarters. The estimation of residual milk served to show that this disproportion did not exist. Similar data were also obtained with other animals. Therefore, in order to exclude any error in the estimation of the development of the quarter of the udder as one of the most important indices characterizing the suitability of cows to machine milking it is necessary to take into account not only milk yield but also residual milk. This may be done in the following way: In first-calf cows being milked by machine three times a day, on the 60th day of lactation at the morning milking milk is drawn from the quarters into a four-chambered apparatus. Then the cow is injected with 18–20 IU oxytocin in the jugular vein and the residual milk is stripped by hand into a four-chambered pail. Similar estimations are done at the noon milking on the 63rd day of lactation and at the evening milking on the 66th day; with twice-a-day milking the test is done on the 60th day of lactation at the morning milking and on the 63rd day at the evening milking. The development of the quarters of the udder is estimated on the basis of these data. This method of evaluation of the quarters helps to evaluate precisely the quality of the daughters of the bull as well as the quality of the udder of the family. This should help in building a herd of cows that conform to the requirements of machine milking.

Bibliography

1. Anonymous. 1953. *Technik für Bauern und Bartner,* **2**: 33–35.
2. Anonymous. 1964. Rekomendatsii po mashinnomu doeniyu korov (Recommendations on Machine Milking of Cows). Ministry of Agriculture, USSR, Moscow.
3. Anonymous. 1965. Rekomendatsii po otsenke vymeni i molokootdachi korov molochnykh i molochno-myasnykh porod (Recommendations on the Evaluation of Udder and Milk Letdown of Cows of Milk and Milk-fat Breeds). Ministry of Agriculture, USSR, Moscow.
4. Anonymous. 1961. Doilnaya ustanovka UDDE–16; Ystroistvo, montazh, eksplutatsiya, ukhod (Milking Installation UDDE–16; Layout, Letdown, Use and Maintenance). Izhevsk.
5. Akulinicheva, M.L. 1959. *Bulleten' Nauchno-tekhnicheskoi Informatsii Vsesoyuznogo N.–I. Instituta Zhivotnovodstvo,* **1**: 3–8.
6. Andreae, U. 1957. *Technic und Landwirtschaft,* Jrg. 9, **7**: 162–164.
7. Andros, V.A. and V.A. Radko. 1965. Reviewers. Sel'skoe khozyaistvo za rubezhom (Agriculture Abroad). *Zhivotnovodstvo,* No. 1; 72–73, report.
8. Anikin, I.V. 1956a. *Molochnoe i Myasnoe Zhivotnovodstvo,* No. 8: 47–50.
9. Anikin, I.V. 1956b. *Nauka i Peredovoi Opyt v Sel'skom Khozyaistve,* No. 9: 27–28.
10. Azimov, G.I. 1955a. Vtoroe soveshchanie po fiziologii s.–kh. zhivotnykh. Tezisy dokladov (The Second Conference on Physiology of Farm Animals. Abstract of Papers). Moscow-Leningrad. pp. 3–4.
11. Azimov, G.I. 1955b. *Zhurnal Obshchei Biologii,* **16**(4): 249–262.
12. Azimov, G.I. 1960. Kak obrazuetsya moloko (How is Milk Secreted?). Moscow.
13. Azimov, G.I. et al. 1961. *Zhivotnovodstvo,* No. 1: 40–48.
14. Azimov, G.I. 1961. XV Mezhdunarodnyi kongress po molochnomu delu (XV International Dairy Congress). Moscow. pp. 319–320.
15. Azimov, G.I. 1961. *Skotovodstvo,* Moscow. t. **2**, pp. 159–193.
16. Azimov, G.I. et al. 1962. *Zhurnal Obshchei Biologii,* **23**(3): 237–238.

17. BARYSHNIKOV, I.A. et al. 1951. *Zhurnal Obshchei Biologii,* **12**(6): 423–439.
18. BARYSHNIKOV, I.A. and G.B. TVERSKOI. 1962. *Physiologicheskii Zhurnal SSSR im. I.M. Sechenova,* **48**(2): 235–238.
19. BARYSHNIKOV, I.A. 1962. *Physiologicheskii Zhurnal SSSR im. I.M. Sechenova,* **48**(4): 373–377.
20. BARYSHNIKOV, I.A. 1963. Materialy Vsesoyuznogo soveshchaniya po teoreticheskim osnovam povysheniya produktivnosti s.–kh. zhivotnykh (Material of the All-Union Conference on Theoretical Principles of Increasing Productivity of Farm Animals). Borovsk, pp. 15–17.
21. BAXTER, E.S. et al. 1950. *The Journal of Dairy Research.* **17**(2): 117–127.
22. BEAL, D.W. 1957. *Farm Mechanization.* **9**(93) 100.
23. BECK, G.H. et al. 1951. *Journal of Dairy Science,* **34**(6): 598–605.
24. BELUGINA, O.P. 1961. *Trudy Vsesoyuznogo s.–kh. Instituta Zaochnogo Obrazovaniya.* Issue No. 4: 60–63.
25. BELUGINA, O.P. and A.F. ORLOV. 1963. Tezisy Dokladov Nauchnoi Konferentsii po Voprosam Fiziologii i Biokhimii s.–kh. Zhivotnykh (Abstracts of Papers Submitted at the State Conference on the Problem of Physiology and Biochemistry of Farm Animals). Baisogala. pp. 23–24.
26. BELYAEVSKII, U.I. 1960. *Zhivotnovodstvo.* No. 9: pp. 74–76.
27. BERKOVICH, E.M. and V.I. TRETEVICH. 1959. Vsesoyuznoe soveshchanie po fiziologii i Biokhimii s.–kh. zhivotnykh, Tezisy Dokladov (All-Union Conference on Physiology and Biochemistry of Farm Animals. Abstract of Papers). Moscow-Leningrad. pp. 9–10.
28. BERZIN, T. 1964. *Biokhimiya Goromonov* (Hormone Biochemistry). Moscow.
29. BILEK, J. and M. JANOVSKY. 1959. *Zivocisna Vyroba* (Animal Husbandry). **9**, 677-682.
30. BOGDASHEV, N.F. and A.P. ELISEEV. 1957. Molochnye zhelezy s.–kh. zhivotnykh (Milk Glands of Farm Animals). Moscow-Leningrad.
31. BORSUK, V.N. and M.G. ZAKS. 1955. *Trudy Instituta Fiziologii im. I.P. Pavlova,* **4**, pp. 81–92.
32. BORSUK, V.N. 1955a. *Trudy Instituta Fiziologii im. I.P. Pavlova,* **4**, pp. 198–212.
33. BORSUK, V.N. 1955b. *Trudy Instituta Fiziologii im. I.P. Pavlova,* **4**, pp. 213–216.
34. BORSUK, V.N. 1957. Voprosy fiziologii s.–kh. zhivotnykh (Problems in Physiology of Farm Animals). Moscow-Leningrad. pp. 293–302.
35. BRAUDE, R. and K.G. MITCHELL. 1952. *The Journal of Endocrinology.* **8**(3): 238–241.

36. BRESLAV, I.S. 1957. Voprosy fiziologii s.–kh. zhivotnykh (Problems in Physiology of Farm Animals). Moscow-Leningrad. pp. 356–359.
37. CARLISLE, D.B. and P.M. JENKIN. 1959. *Nature,* **183**(4656): 336–337.
38. CARUOLO et al. 1955. *Journal of Dairy Science.* **38**(11): 1283–1292.
39. CHERKASHCHENKO, I.I. 1957. *Bulleten Nauchno-tekhnicheskoi Informatsii Vsesoyuzonogo n.–i. Instituta Zhivotnovodstva. Aspirantskii.* pp. 8–10.
40. CLOUGH, P.A. and F.H. DODD. 1956. *Agriculture,* **63**(7): 334–335.
41. COMBERG, G. and H.G. ZSCHOMBLER. 1959. *Archiv für Tierzucht.* **2**(1): 29–70.
42. COWIE, A.T. and J.S. TINDALL. 1961. *The Journal of Endocrinology,* **23**(1): 79–96.
43. CUBA, S. and F. CZAKO. 1960. *Allattenyesztes,* **9**(2): 105–108.
44. DENAMUR, R. and I. MARTINET. 1961. *Annales d'Endocrinologie,* **22** (5): 776.
45. DIDKOVSKAYA, M.K. 1955. *Trudy Odesskogo s.–kh. Instituta,* **7**: 149–158.
46. DIETRICH, K. 1958a. *Tierzucht,* Jrg. 12. **5**: 146–153.
47. DIETRICH, K. 1958b. *Die Deutsche Landwirtschaft,* Jrg. **9**, 9: 445–446.
48. DODD, F.H. and A.S. FOOT. 1947. *The Journal of Dairy Research,* **15** (1): 1–17.
49. DODD, F.H. and A.S. FOOT. 1948. *Agriculture,* **55**(6): 238–242.
50. DODD, F.H. et al. 1949. *The Journal of Dairy Research,* **16**(3): 301–309.
51. DODD, F.H. and A.S. FOOT. 1953. *The Journal of Dairy Research,* **20**(2): 138–145.
52. DODD, F.H. and P.A. CLOUGH. 1962. XVI International Dairy Congress, Copenhagen, Vol. **A**: 89–97.
53. DONGIDOO, E. 1962. Fiziologiya s.–kh. Zhivotnykh (Physiology of Farm Animals). Moscow. pp. 348–358.
54. DONKER, J.D. et al. 1954a. *Journal of Dairy Science,* **37**(11): 1261–1268.
55. DONKER, J.D. et al. 1954b. *Science,* 119 (3080), 67–68.
56. DUSEMBIN, KH. 1957. Voprosy fiziologii s.–kh. zhivotnykh (Problems in Physiology of Farm Animals). Moscow-Leningrad. pp. 316–324.
57. EISENREICH, L. 1954. *Milwissenschaft,* Jrg. **9**, 12: 398–402.
58. EISENREICH et al. 1955. *Landtechnik,* Jrg. **10**, H. 12: 508–509.
59. ELLIOTT, C.M. 1959. *Dairy Science Abstracts,* **21**(11): 481–490.
60. ELLIOTT, C.M. 1961. *The Journal of Dairy Research,* **28**(2): 123–129.
61. ESPE, D. 1950. Sekretsiya moloka (Milk Secretions). Moscow.

62. Fedii, E.M. 1961. *Zhurnal Obshchei Biologii,* **22**(1): 75–77.
63. Fitzpatrick, R.J. 1961. Oxytocin, *Proc. Intern. Symp. Montevideo,* 1959, Oxford. pp. 358–377.
64. Gar'kavyi, F.L. 1965. *Zhivotnovodstvo.* No. 2: 29–33.
65. Gedde-Dahl, T. 1961. *Tidsskrift for det Norske Landbruk,* Arg. **68**. H. 7: 191–197.
66. Gorodetskaya, T.K. 1962. Fiziologiya s.–kh. zhivotnykh (Physiology of Farm Animals). Moscow, pp. 313–327.
67. Goto, T. et al. 1959. *Bulletin of the National Institute of Agricultural Sciences, Series G, Animal Husbandry,* Japan, No. 18: 163–168.
68. Grachev, I.I. 1252a. *Doklady Akademii Nauk, USSR,* **84**(2): 397–400.
69. Grachev, I.I. 1952b. *Doklady Akademii Nauk, USSR,* **86**(2): 441–444.
70. Grachev, I.I. 1954. *Uchenye Zapiski Leningradskii Gosudarstvennyi Universitet. Seriya Biologicheskikh Nauk,* Vol. 176, No. 37, pp. 164–170.
71. Grachev, I.I. 1956. *Vestnik Leningradskogo Universiteta,* Vol. 9, *seriya biologii,* No. 2: 85–90.
72. Grachev, I.I. 1959. Vsesoyuznoe soveshchanie po fiziologii i biokhimii s.–kh. zhivotnykh. Tezisy dokladov (All-Union Conference on Physiology and Biochemistry of Farm Animals. Abstract of Papers). Moscow-Leningrad. pp. 250–251.
73. Griffin, T.K. and F.H. Dodd. 1962. *The Journal of Dairy Research,* **29**(2): 207–210.
74. Grigor'eva. N.I. 1956. *Doklady TSKhA.* No. 22: 512–518.
75. Gumenyuk, I.G. and S.P. Vasina. 1958. *Sbornik Trudov Penzenskogo Instituta,* No. 2: 325–339.
76. Hammond, J. 1913. *Quarterly Journal of Experimental Physiology,* **6**(4): 311–338.
77. Happ, H. 1961. *Der Tierzuchter,* Jrd. **13**, No. 17, pp. 432–434.
78. Hawker, R.W. and V.S. Roberts. 1957. *The British Veterinary Journal,* **113**(11): 459–464.
79. Heller, H. 1961. Oxytocin. *Proc. Intern. Symp. Montevideo,* 1959. Oxford, pp. 3–22.
80. Henderson, G.O. and P.M. Rivz. 1957. Kormlenie i Soderzhanie Molochnogo Skota (Feeding and Management of Dairy Animals). Moscow.
81. Hooven, N.W. and R.D. Plowman. 1961. *Journal of Dairy Science,* **44**(6): 1199.
82. Ioannisyan, S.L. 1959. Dostizheniya nauki v zhivotn avodstve (Achievements of Science in Cattle Breeding). Doklady na Yubileinoi Sessii Akademii Nauk SSSR i VASKhNIL, Posvyashchennoi

100-letiyu so dnya Rozhdeniya I.V. Michurina. Moscow, pp. 228–239.

83. Johansson, I. 1952. *Acta Agriculture Scandinavica*, **2**(1): 82–102.
84. Johansson, I. and P. Malven. 1960. *Zeitschrift für Tierzüchtung und Züchtungsbiologie*, Bd. **74**, H. 1, pp. 1–13.
85. Jorgensen, K.E. and E.V. Garulolo. 1963. *Journal of Dairy Science*, **46**(6): 624.
86. Kaveshnikova, K.I. 1960a. *Zhivotnovodstvo*, No. 7: 83–87.
87. Kaveshnikova, K.I. 1960b. *Vestnik s.-kh. Nauki*, No. 8: 136–138.
88. Kaveshnikova, K.I. 1963a. *Fiziologicheskii Zhurnal SSSR im. I.M. Sechenova*, **49**(1): 101–110.
89. Kaveshnikova, K.I. 1963b. Tezisy dokladov nauchnoi konferentsii po voprosam fiziologii i biokhimii s.-kh. zhivotnykh (Abstracts of Papers Submitted at the State Conference on the Problem of Physiology and Biochemistry of Farm Animals). Baisogola, p. 52.
90. Kaveshnikova, K.I. 1964. Fiziologicheskie mekhanizmy mashinnogo doeniya (Physiological Mechanisms of Machine Milking). Moscow–Leningrad. pp. 20–32.
91. King, N. 1956. Obolochi Zhirovykh Sharikov Moloka (Fat Globules of Milk). Moscow.
92. Khazak, S. and Bernet, G. 1961. *Doklady TSKhA*. No. 65: pp. 15–27.
93. Klochko, I.M. 1954. Razdoi korov v kolkhozakh i sovkhozakh Ukrainy (Milking of Cows on Kolkhozes and Sovkhozes in Ukraine). Moscow.
94. Knodt, C.B. et al. 1947. *Journal of Dairy Science*, **30**(8): 580–581.
95. Kobzev, V.S. 1955. Sbornik dokladov tretego vsesoyuznogo soveshchaniya po molochnomu delu (Collection of Papers Presented at the Third All-Union Dairy Conference). Moscow. pp. 55–58.
96. Kokorina, E.P. 1956. *Doklad Akademii Nauk, SSSR*, **108**(4): 746–749.
97. Kokorina, E.P. 1957. Voprosy fiziologii s.-kh. zhivotnykh (Problems in Physiology of Farm Animals). Moscow-Leningrad. pp. 44–49.
98. Korolev, V.F. 1960a. *Zhivotnovodstvo*. No. 1: 38–43.
99. Korolev, V.F. 1960b. *Zhivotnovodstvo*. No. 9: 71–74.
100. Korolev, V.F. 1962. Doilnye mashiny (Milking Machines). Moscow.
101. Koshi, J.H. and W.E. Petersen. 1955. *Journal of Dairy Science*, **38** (7): 788–796.
102. Kravchuk, A. 1965. *Molochnoe i Myasnoe Skotovodstvo*. No. 7, 39–40.
103. Kreilis, M.L. 1964. Fiziologicheskie mekhanizmy mashinnogo

doeniya (Physiological Mechanisms of Machine Milking). Moscow-Leningrad. pp. 52–59.

104. Kudryashov, G.V. 1954. *Zhivotnovodstvo*. No. 12: 86–90.
105. Kuznetsov, S.V. (Reviewer). 1965. Sel'skoe khozyaistvo za rubezhom (Agriculture Abroad). *Zhivotnovodstvo*. No. 4: 66–67.
106. La Bella, F.S. et al. 1963. *Archives of Biochemistry and Biophysics*, **100**(3): 399–408.
107. Lakshmanan, S. et al. 1958. *Journal of Dairy Science*, **41**(11): 1601–1608.
108. Larsen, L.G. and G.V. Eskedal. 1957. Kormleniye vysokoproduktivnykh molochnykh korov (Feeding High-yield Milch Cows). Moscow.
109. Lascelles, A.K. 1961. *Nature*, **191**(4796): 1404–1405.
110. Lascelles, A.K. 1962. *Quarterly Journal of Experimental Physiology and Cognate Medical Sciences*, **47**(1): 48–56.
111. Lauener, F. 1959. *Zentralblatt für Veterinarmedizin*, Bd. **6**, H. 9, pp. 832–853.
112. Leontovich et al. 1932. *Elektrifikatsiya Sel'skogo khozyaistva*, No. 5–6: 48–63.
113. Lightbody, L.G. and A.J.W. Murray. 1960. *The Queensland Journal of Agricultural Science*, **17**(3): 155–163.
114. Luick, J.R. 1961. *Journal of Dairy Science*, **44**(4): 652–657.
115. Makarov, A.P. 1960. Povyshenie proizvoditel'nosti truda v zhivodnovodstve i snizheniye sebestoimosti produktsii (Raising the productivity of labor in animal husbandry and lowering the cost of production). *Trudy Vtoroi Nauchnoi Konferentsii*. Krasnodar, pp. 347–354.
116. Manunta, G. and A. Marongiu. 1961. *Bollettino della societa Italiana di biologia sperimentale*, **37**(11): 510–512.
117. Marchenko, G.M. 1957. *Trudy Kubanskogo s.-kh. Instituta*, Krasnodar, No. **3**(31): 89–100.
118. Marchenko, G.M. 1959. Vsesoyuznoe soveshchaniya po fiziologii i biokhimii sel'skokhozyaistvennykh zhivotnykh Tezisy dokladov (All-Union Conference on Physiology and Biochemistry of Farm Animals. Abstract of Papers). Moscow-Leningrad. pp. 269–270.
119. Marchenko, G.M. 1960. Voprosy sekretsii moloka i molokootdachi y korov (Problems of secretion and letdown of milk). Dissertatsiya. Krasnodar. 1960.
120. Martynov, V. 1913. Nervnye apparaty v soske cheloveka i mlekopitayushchikh (Nervous apparatus in teats of human beings and mammals). Dissertatsiya. St. Petersburg.
121. Martyugin, D.D. 1941. *Trudy Moskovskoi s.-kh. Akademii im. K.A. Timiryazeva*. Moscow. Vol. 5, No. **3**, pp. 119–150.

122. Martyugin, D.D. 1944. *Trudy TSKhA*. No. 31, pp. 149–184.
123. Martyugin, D.D. 1950. *Sovetskaya Zootekhniya*. No. 5: 75–85.
124. Martyugin, D.D. 1952. *Sotsialisticheskoe Zhivotnovodstvo*, No. 4: 68–72.
125. Martyugin, D.D. 1954a. *Zhivotnovodstvo*, No. 2: 56–58.
126. Martyugin, D.D. 1954b. *Izvestiya TSKhA*. No. 3: 129–148.
127. Martyugin, D.D. 1955. *Izvestiya TSKhA*. No. 2: 225–238.
128. Meshchaninov, S.I. 1952. Osnovnye izmeneniya vymeni tagilo-ostfrizkogo skota v svyazi s massazhem, doeniem, vozrastom i produktivnostyu (Principal changes in the udder of tagilo east frierian cattle in relation to massage, milking, growth and productivity). Dissertatsiya, Moscow. VIZh.
129. Mokeev, A.E. 1955. *Zhurnal Obshchei Biologii*, **16**(2): 156–164.
130. Moroz, V. 1965. *Molochnoe i Myasnoe Skotovodstvo*. No. 6: 23–25.
131. Murray, A.J.W. and L.G. Lightbody. 1962. *Queensland Journal of Agricultural Science*, **19**(2): 255–265.
132. Myand, A.E. et al. 1964. Mekhanizirovannaya ferma v 'Kutuzovke' (Mechanized Farm in 'Kutuzovka'). Moscow.
133. Nazarenko, V.I. (Reviewer). 1965. Sel'skoe khozyaistvo za rebuzhom (Agriculture Abroad). *Zhivotnovodstvo*. No. 5: 51–53, Abstract.
134. Nekrasova, M.A. 1955. Sbornik rabot Leningradskogo veterinarnogo instituta (Collection of Works of the Leningrad Veterinary Institute). No. 15: 142–145.
135. Novikov, A. 1962. *Mezhdunarodnyi Sel'skokhozyaistvennyi Zhurnal*, No. 6: 93–97.
136. Olenev, V. 1965. *Molochnoe i Myasnoe Skotovodstvo*, No. 2 (43–45).
137. Olenov, Yu. M. 1954. *Doklad Akademii Nauk SSSR*, **97**(2): 361–364.
138. Orlov, A.F. 1954. *Fiziologicheskii Zhurnal SSSR im. I.M. Sechenova*, **40**(1): 39–44.
139. Orlov, A.V. 1956. Izmenenie kachestva moloka v protsesse doeniya korov raznymi sposobami (Changes in the quality of milk of cows milked by different appliances). Avtoreferat dissertatsii. Moscow.
140. Pavlov, I.P. 1951. Dvadtsatiletnii opyt ob'ektivnogo izucheniya vysshei nervnoi deyatel'nosti (povedeniya) zhivotnykh [Twelve-year Experiment of Objective Studies on the Influence of the Higher Nervous System (Behavior) of Animals]. Izd. 7e, Moscow.
141. Peinovich, M.L. 1957. Novye dannye po fiziologii doeniya (New data on the physiology of milking). *Trudy Sibirskogo N.–I. Instituta Zhivotnovodstva*. No. 13.
142. Peinovich, M.L. 1962. Mashinnoe doeniye korov (Machine Milking of Cows). Moscow, pp. 149–155.
143. Petersen, W.E. 1945. *Hoard's Dairyman*, **90**(21): 619, 626, 627, 633.

144. PETIT, K.L. 1961. *Mededelingen van de landbouwhogeschool en de opzoekingsstations van de staat te gent,* Deel **26**, No. 1: 215–258.
145. PITERMAN, YA. A. Mashinnoe doeniya korov (Machine Milking of Cows). Moscow. pp. 136–148.
146. POLUPANOV, E.P. 1964. Mashinnoe doeniye korov (Machine Milking of Cows). Moscow. pp. 43–56.
147. POPOVICH, D. 1958. *Doklady Akademii Nauk SSSR.* **121**(1): 186–189.
148. POUTOUS, M. 1962. *La revue de l'élevage,* **17**, an., 32 sp. num., pp. 65–70.
149. RICHARDSON, K.C. 1950. *Proceedings of the Royal Society of London,* Series B. Vol. **136**, pp. 35–45.
150. ROSTOVSHCHIKOVA, T.M. 1962. *Trudy Permskogo s.-kh. Instituta,* **18**(4): 255–259.
151. RYAZANKIN, K.G. 1962. Fiziologiya s.–kh. zhivotnykh (Physiology of Farm Animals). Moscow. pp. 265–279.
152. SALMIN, I.P. 1956. XI Konferentsiya filiala yuga RSFSR vsesoyuznogo obshchestva fiziologov, Biokhimikov i farmakologov. Tezisy Dokladov (XI Conference of the Southern RSFSR Section of the All-Union Society of Physiologists, Biochemists and Pharmocologists. Abstract of Papers). Stavropol'. pp. 225–228.
153. SCHMIDT, G.H. et al. 1964. *Journal of Dairy Science,* **47**(2): 152–155.
154. SCHNERCH, F. 1959. *Archiv für Tierzucht,* Bd. **2**, H. 1, pp. 71–85.
155. SEMENOVYKH, N.E. 1965. *Veterinariya,* No. 1, pp. 92–95.
156. SEMJAN, S. 1962. *XVI International Dairy Congress,* Kobenhavn, Vol. A, pp. 17–24.
157. SENTOVSKI, J. 1960. *Nowe Roinictvo,* R.G., No. 24, pp. 28–30.
158. SHAW, J.C. 1942. *Journal of Dairy Science,* **25**(12): 1051–1055.
159. SHI CHZHUN' YUAN'. 1962. *Byulleten' Eksperimental'noi Biologii i Meditsiny,* No. 12: 22–26.
160. SIMS, J.A. 1962. *Iowa Farm Science,* **17**(6): 15–16.
161. SINESHCHEKOV, A.D. and Z.I. SHEREMET. 1953a. *Fiziologiya Pitaniya S.–Kh. Zhivotnykh,* Moscow. pp. 24–58.
162. SINESHCHEKOV, A.D. and Z.I. SHEREMET. 1953b. *Fiziologiya Pitaniya S.–Kh. Zhivotnykh,* Moscow. pp. 280–287.
163. SINESHCHEKOV, A.D. 1956. Fiziologiya pitaniya i rezhim dnya s.–kh. zhivotnykh (Physiology of Nutrition and Dairy Regime of Farm Animals). Moscow.
164. SINYAVS'KII. V.N. 1960. Sotsialistichne tvarinnitstvo, No. 2, pp. 39–40.
165. SMITH, V.R. 1947. *Journal of Dairy Science,* **30**(9): 703–715.
166. SMITH, V.R. and W.E. PETERSEN. 1948. *Journal of Dairy Science,* **31** (7): 589–593.